No. 2058
$14.95

THE CAR AND ITS WHEELS—a Guide to Modern Suspension Systems
by JAN P. NORBYE

MODERN AUTOMOTIVE SERIES

TAB BOOKS Inc.

BLUE RIDGE SUMMIT, PA. 17214

TX 534-927

FIRST EDITION

FIRST PRINTING—JUNE 1980

Copyright © 1980 by TAB BOOKS Inc.

Printed in the United States of America

Library of Congress Cataloging in Publication Data

Norbye, Jan P.
    The car and its wheels.

    Includes index.
    1. Automobiles—Springs and suspension. I. Title.
TL257.N67   629.2'43   79-23615
ISBN 0-8306-9719-5
ISBN 0-8306-2058-3 pbk.

Cover photo courtesy of Mercedes-Benz.

Other TAB Books by the author:

No. 2004   *Chassis Tuning*
No. 2037   *Streamlining and Car Aerodynamics*
No. 2046   *Modern Diesel Cars*
No. 2052   *The Complete Handbook of Front Wheel Drive Cars*
No. 2069   *The Complete Handbook of Automotive Power Trains*

Contents

Introduction

Wheels—the very word is often synonymous with automobile. When a fellow asks, "You got wheels?" he is not inquiring about a stack of wheels in a warehouse but simply trying to find out if his interlocutor has a car. When a famous author needed a title for a novel about Detroit, he did not call it *Cars,* but *Wheels*. It means the same thing. American slang has intuitively seized the importance of wheels. It is no exaggeration to say that a car is only as good as its wheels.

The statement is incomplete, however, for wheels are useless without a suspension system. What links the wheels to the car is what this book is all about. We will look at the wheels, themselves, and will discuss chassis design and layout.

Our starting point is the wheels, and to be logical, nothing can reasonably be said about wheels until we have defined their duties. Any child will tell you the duty of the wheel is to roll. That's perfectly true, but it is not the whole truth. It is a little more complicated than that.

Jan P. Norbye

Chapter 1

The Duties of the Wheel

The wheel, in the sense that we are using the term here, is an assembly made up of a hub, a wheel, and a tire. Four of these assemblies assure the car's contact with the road, its only such contact. Every aspect of running, turning, starting, and stopping is a function of action by the wheels (Fig. 1-1). And there's more. The wheels must carry the weight of the car (and its payload). Well, what does that mean? It means that the wheel hubs and suspension system must be rugged enough to handle such load and not collapse. It's not just the static load, but also dynamic load variations due to changes in the car's speed and direction.

All wheels must have brakes. Some have to steer and others have to drive; some even do both. This means that the wheels must be linked to the car by additional mechanisms separate from those of the suspension system.

Braking and driving are the simplest, for they affect only the most basic function of the wheel—that of rolling on the roadway. It's a one-dimensional concept: to roll or not to roll. Of course, it is of vital importance that the wheels be matched to the performance of the car in terms of size and speed capacity, and we'll go deeper into that later on.

But steering is something else. Giving a wheel freedom to steer involves another dimension (Fig. 1-2). Namely, that of turning the hub on a more or less upright axis, so as to make the wheel roll in a different direction. That means making basic changes in the suspension system as well as providing a steering linkage.

Whether wheels that steer also have to drive (as with front-wheel-drive cars) is not a matter that seriously alters the situation. Only the suspension system and steering linkage must be so designed that their parts don't get in the way of the drive shafts.

Fig. 1-1. The wheels dictate all dynamic behavior in the car—speeding up and slowing down, setting the course and changing it.

Suspension Permits Vertical Travel

Several times now we have mentioned suspension systems, without explaining what they are and why cars must have them. A wheelbarrow does not have a suspension system. Its single wheel turns on a spindle whose ends are rigidly anchored in a fork that is part of the frame. On a cart, the two wheels are usually mounted on a rigid axle whose casing is bolted to the cart frame. That is not a suspension system.

Suspension systems provide flexibility for the wheels relative to the vehicle frame and body. And for any four-wheeled vehicle, that is essential. Why? Because road surfaces are never perfect. Even the smoothest roadway is a succession of bumps, ridges and depressions. If the four wheels were rigidly attached to the chassis, one wheel would be lifted off the ground at all times, each one taking its turn in a random pattern, so that the vehicle, in fact, would never be supported by more than three wheels at any time.

In other words, suspension systems make it possible for the wheels to move up and down without affecting the vehicle frame. This is a deflection which the engineers call vertical travel, wheel travel, or just travel.

Let's look more closely at this four-wheeled machine that has only three wheels on the ground as it moves along the roadway. It is not impossible for such a car to work. There is always one wheel on the ground that drives, and always at least one that steers; and usually, that's enough.

But look what happens to the people in the car! They are victims of intollerable jolts, shaking and vibrations. Nobody could stand to ride very far or very fast in such a car. Goods carried by it could be damaged. And damage would certainly be done to the vehicle itself. It would crack up and break into pieces after a ridiculously short mileage.

It is the duty of the suspension to provide ride comfort for the occupants of the car and to protect the vehicle itself from the shocks that the wheels encounter along the roadway. We mentioned flexibility as being a characteristic of the suspension system. That introduces the idea of a spring. Yes, springing is a key function of the suspension system. All suspension systems contain a flexible component that regulates the wheel travel. The other key function of the suspension system is to locate the wheels, in other words, to restrict their freedom of movement to the intended planes and to the desired extent only.

It is good to separate in our minds the ideas of flexing or springing, on one hand, and location of the wheels on the other, even if, as in some suspension systems, the two functions are combined in the same physical parts. In modern cars, however, springs are divorced from wheel-locating duties, and their wheels remain attached even if the springs are removed.

Springs have a difficult task. They must support the static load of the vehicle and its payload. They must flex as the car runs along the road, offering the wheels enough travel to assure ride comfort but without such looseness as to produce seasickness in the passengers.

Fig. 1-2. One wheel may be a small part of the car, but in cornering situations, a single-wheel—and its tire—makes all the difference between keeping directional control and losing it.

They should absorb the irregularities of the road surface as much as possible. If the spring is too stiff, bumps will not be fully absorbed, and the shock will be partly transmitted to the vehicle structure and interior. If the springs are too soft, they will absorb the bumps, but not without risk of compromising the car's road-holding ability and dynamic stability.

While soft springs and long wheel travel are considered essential for proper ride comfort, these requirements tend to conflict with the conditions needed for good road-holding for, as a basic rule, handling precision varies inversely with wheel deflection. In other words, some road-holding ability is lost whenever one or more wheels move up or down from normal position, and the loss is greater the further the wheel has been displaced (Fig. 1-3). On the other hand, overly stiff springs can also be detrimental to good road-holding, because their failure to flex sufficiently can cause loss of contact between the tire and the road surface.

Spring stiffness, therefore, is selected to give the best combination of requirements that, within the limits of normal practice tend to oppose each other (See Fig. 1-4). Here it must also be taken into account that some spring action is performed by the tire. Properly inflated tires will absorb

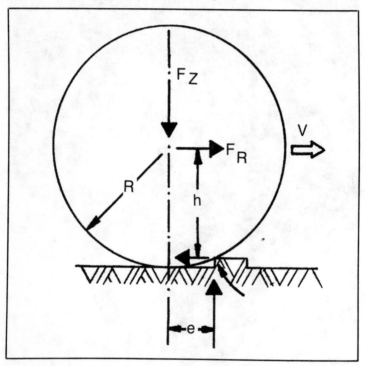

Fig. 1-3. The extent of wheel travel is the same for all wheels on a given bump, but the rate of wheel travel is more gradual for bigger wheels because distance is longer.

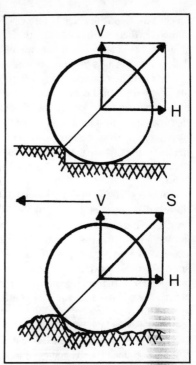

Fig. 1-4. Bumps not only push the wheel up, but also exert a force against its direction of travel. The total impact force S breaks down into a vertical component V and a horizontal component H.

minor surface unevenness, but larger irregularities will deflect the whole wheel (Fig. 1-5).

Good Reasons for Smaller Wheels

Wheel and tire size play an important part in this connection. A small wheel falls deeper into a pothole than a large one (because the bigger tire

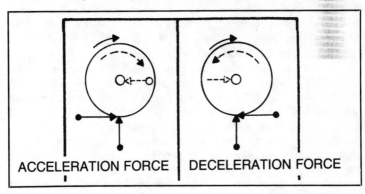

ACCELERATION FORCE   DECELERATION FORCE

Fig. 1-5. Acceleration and braking thrust forces act through the wheels. Wheels must be suspended so that they can handle these forces without losing their aim.

touches the opposite ridge before reaching bottom and, thus, bridges the pothole). When meeting a bump, the larger tire makes contact first that is, at a longer distance from the wheel hub. Even if the large wheel is deflected as much as the small one hitting the same bump, the large wheel climbs over the obstacle along a flatter gradient. That means the shock load passed on to the suspension system is reduced. The spring is deflected more gradually when the wheel and tire are of large diameter. And that translates into improved ride comfort.

Large tires tend to run cooler because they make fewer revolutions per mile, and that means a reduced number of flexing cycles for any given distance. Flexing heats up the tires, and overheating can lead to tire failure.

A quick look at the cars around us is enough to tell you that, despite these advantages of large wheels, wheels have in fact been getting smaller. There are valid reasons for this, and it's not because roads are getting smoother. It is, in part, because cars are getting smaller. Small wheels take up less space. That means wheel housings can be made smaller so that there is less intrusion into the interior. Small cars with large wheels lose front leg room (floorboard area and lack of footrest space around the pedals) and rear seat width, as well as rear doorway opening area on four-door models.

Small wheels tend to lower the whole chassis, which brings the center of gravity closer to the ground, and that makes the car more stable. If it also reduces the ground clearance, that is not regarded as a serious objection. Small wheels are fitted with smaller tires, and that means lower cost for the manufacturer. That's a very important reason for the trend to smaller wheels. It is true that smaller wheels impose limits on brake disc and drum diameter. Again, that is not a problem because modern disc brakes offer adequate braking power even with quite small sizes.

Finally, small wheels are lighter than big ones. And that's very important, for the wheel is counted as unsprung weight. The sprung weight is everything that's on the other side of the springs and getting the benefit of spring action. Unsprung weight is the rest—the wheels and all the stuff that moves up and down with wheel deflections.

The greater the unsprung weight, the greater the disturbance in the car caused by roadway unevenness, and the more handling precision which is lost (Figs. 1-6 and 1-7). At the same time, ride comfort deteriorates and noise level is increased, with stronger vibrations, shock transfer and bending stress levels in the whole vehicle structure. With low unsprung weight, tires are better able to keep in continuous contact with the road surface and, thereby, minimize any partial loss of traction and sidebite, making for better road-holding.

Traction, in this context, refers to both driving and braking torque and their effectiveness against the road surface. Sidebite is a non-technical term for lateral adhesion and applies equally to both steered and non-steered wheels. Taken together, traction and sidebite spell road-holding.

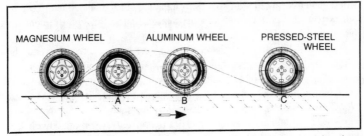

Fig. 1-6. Lightweight wheels regain contact with the road surface quicker than heavier wheels. Magnesium wheel returns at A, aluminum wheel at B, and pressed-steel wheel at C.

In connection with spring stiffness, we have also used the term handling precision. No need to explain what precision means. But handling, as the term applies to cars, is a much-misunderstood notion.

Good Handling Means Well-Behaved

That a car handles well does not mean that it is easy to handle, that is, requires little effort to turn the steering wheel when parking. It does not mean that its exact size is easily seen or guessed from the driver's seat

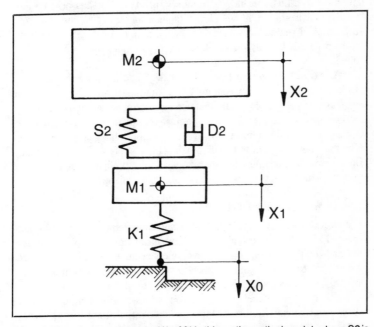

Fig. 1-7. The wheel is represented by M1 in this mathematical model, where S2 is the suspension spring and D2 the shock absorber, K1 expresses the springing effect in the tire, and M2 the sprung mass (load on one wheel).

when maneuvering in tight spaces. It means that its road behavior corresponds to the driver's intentions as communicated through the controls (steering wheel, accelerator and brakes).

A car's handling precision depends on many factors beyond the wheels and suspension systems, such as front/rear weight distribution, center of gravity height, and polar moment of inertia. For a detailed explanation of these and other car design parameters, see my TAB book No. 2004 *Chassis Tuning.* Here we will only touch briefly on matters of vehicle layout and architecture in order to concentrate on the wheels and how they are attached, with the aim of making clear how suspension systems affect the dynamic behavior of a car.

Car tires, as opposed to motorcycle tires, are designed to run on their full tread width, which means that the wheels they are mounted on are expected to be mounted in an upright position and not tilt far one way or another for any length of time (Fig. 1-8).

The wider the tire tread, the more important it is to avoid tilting the wheel, for the footprint is usually shorter. In other words, the ordinary six-inch-wide rim carries a tire with a footprint of roughly rectangular shape, stretching in the tire's direction of rolling. These tires, running under the load of the cat at prescribed inflation pressure, are not round but flatten out against the road surface. A wider-high-performance tire may have a squared-off footprint, wider and shorter, but not necessarily greater in area. In profile, it is rounder with less flattening against the road surface. And racing tires made for 12-inch to 16-inch wide rims have footprints perhaps only an inch long. From the side, they look perfectly round.

Any tilt of the widest tires will lift the tread on one side, and even slight tilt angles can cause dramatic reductions in footprint area. The suspension systems of the cars that use them must be devised to minimize any changes in such tilt (camber angles). Even with tires of standard width, it is considered desirable to avoid large camber changes. Tires usually give their best traction and sidebite when the wheels are running upright. But, in practice, camber changes are inevitable, though greatly varyiable in degree according to suspension design principles.

Steered Wheels at the Front

Steered wheels have different suspension requirements from non-steered wheels. For normal cars, that means front wheels place different demands on the suspension system than rear wheels. Since we have been going down to basics earlier in this chapter, it is perhaps not out of place to ask why it is always the front wheels that steer.

Steered rear wheels are what you have when backing up. It gives surprising maneuverability, but the car is essentially unstable. Of course, the geometry of the steered wheels is set up to go the other way. Some of the instability could be cured by revising the front wheel alignment for going backwards, so as to obtain a self-centering action.

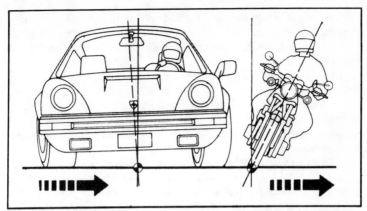

Fig 1-8. Cars lean the wrong way on curves, while the motorcycle balances the side forces. The best the car's suspension system can do for the wheels in a cornering situation is to keep them as upright as possible.

Fork-lift trucks have rear wheel steering because it is the best in their environment. They need the maneuverability to move in narrow corridors inside warehouses and must be able to turn on the proverbial dime. On the other hand, they run at moderate speed over short distances only. But a passenger car made for traveling at highway speeds would have serious control problems if the steering worked on the rear wheels.

With steering on the front wheels, the rear ones are simply tracking, but on a shorter turn radius. Make a turn in firm sand or fresh snow and you'll see four separate and distinct tracks, one for each wheel, with the inside rear track closest to the turn center emerging from and merging back with the inside front track when the car resumes a straight course. Similarly, the outside rear track is laid between the front wheel tracks, but it is closer to the outside front track than the inside since the front and rear wheels on the same side run in the same track when the car is not making a turn. This presumes equal front and rear tracks, which is normal practice (difference in front and rear track is usually kept at less than half the width of the tire tread).

Now, if the rear wheels are steered, the tracks on a curve will present a different picture. It is the rear wheels that make the widest curve, —that is, lay the tracks with the longest turn radius. That's important, for it puts the car into an oversteering attitude, and that introduces a problem about coming out of the turn. There is a lag in response which, in combination with the oversteering attitude of the car, makes it difficult for the driver to know when to straighten out the wheels. It is natural to wait until the car has completed the turn—but then it's too late. The car will continue to turn for a moment. As a result, the driver will have to make corrections by steering the other way (countersteering).

Now, it is a fact of vehicle dynamics that cars which corner in an understeering attitude are inherently stable, and any excess understeer has a self-diminishing tendency. With an oversteering attitude, the car is in an unstable condition and, what's worse, oversteer, once started, tends to feed on itself, posing a progressively stronger threat to the driver's control of the car.

The greater the speed, the longer the distance needed to correct the vehicle attitude, and the more difficult it becomes to make accurate steering corrections. It is natural to crank in a greater steering angle than is actually needed (overshoot), which only reinforces the instability and introduces the need for further correction. That's why cars do not have steered rear wheels.

Nor is four-wheel steering desirable for cars in normal use for the same reasons. All the drawbacks of steered rear wheels are still there, though the driver has an easier task in keeping the vehicle under control due to having steered front wheels as well. Four-wheel steering is used on some military vehicles and off-road recreational vehicles, intended mainly for low-speed operation in difficult terrain where extremely tight turning circles may be necessary to keep the machine from getting stuck.

Alignment Introduces Ackermann Principle

So, from here on, we are talking about ordinary road cars whose front wheels are wheels that steer, with rear wheels that are non-steered. Rear wheels have simple alignment needs. Ideally, they want to stand upright (zero camber), pointing straight to the front wheels (no toe-in or toe-out). Theoretically, their needs can be satisfied by a suspension system that keeps the hub always in the same place.

Front wheels, on the other hand, complex alignment needs. They need a steering axis, which means two attachment points, at different heights, for the hub. A third attachment is needed to control the steering angle. To understand the intricacies of front wheel alignment, we must acquaint ourselves with something known as the Ackermann principle.

To explain it, it is best to start with an old horsedrawn carriage, such as a Conestoga wagon, if you will. It had a front axle with the wheel hubs rigidly attached to its ends. To turn, the whole axle assembly was steered to one side, turning as a unit. This placed the steering to one side, turning as a unit. This placed the steering axis at the center of the axle, equally far from either front wheel. It also put both front wheels at right angles to the turn radius, a prerequisite for the wheel's ability to follow the curve.

On cars, the front axle does not turn. The wheels are mounted on stub axles, each having its own steering axis alongside the hub (think of it as the king-pin, even if king-pins have been almost universally replaced by ball-joints). Each wheel hub also has a steering arm. The length of this steering arm determines the leverage exerted by the steering linkage on the wheel (and plays a part in determining the overall steering ratio). But more important is the angle of the steering arms.

They do not point straight backwards, parallel with the direction of the wheel, but point towards the center of the rear axle, midway between the rear wheels. The two steering arms can be connected to each other via a simple tie rod. When the front wheels point straight ahead, it runs in a straight line, perpendicularly across the car.

In practice, the steering arms are not aimed exactly at the rear axle center, but at an instantaneous center further forward. This variance from what is theoretically right does not disprove the Ackermann principle. It remains valid, but the combined dynamic reactions of the total car have made it advisable to modify the theoretical geometry in order to obtain true theoretical behavior in the suspension and steering systems.

Shortcomings in actual hardware can also cause some deviation from the theoretically ideal front wheel alignment. For instance, the rolling resistance in the front tires acts as a force that tries to push the wheels back. Since the locating members of the suspension system keep them where they are, the force makes the most of any play that exists in the suspension and steering linkages and tends to turn the wheel away from its point of attachment. The result is a tendency to toe out. Since it is impractical to completely eliminate play in the suspension and steering (for reasons ranging from cost to problems of ride comfort, noise and vibration), most cars have their front wheels aligned with some measure of static toe-in so that they will run straight at speed.

On most cars, the front wheels are also aligned with a slight positive camber angle, which means they lean out. Negative camber is an inward tilt. The reason is that the weight of the car tends to compress the springs, and that leads to taking up any play in the suspension system. With most suspension systems, that tends to force both front wheels into an inward tilt. It is to counteract that process that a static positive camber angle is usually chosen.

A secondary reason is a desire to counterbalance the toe-in. Since the wheel tends to turn in the direction it is tilted, a slight positive camber angle can cancel out the toe-in.

A cambered wheel follows a circular path about a point that is located at the intersection of the wheel axis with the road surface. (See Fig. 1-9).

Technology Diminishes Ackermann

In the last twenty years, there has been a trend away from static toe-in and positive camber, paired with a growing disregard of the Ackermann principle. The main reason for this is the progress made in tire technology during this period. By that is meant not only improvement in tire construction and materials, but also our understanding of the dynamic behavior of tires.

Just as tires flex a bit when hitting roadway unevenness and assist the springs in absorbing the shock loads, the tires have a certain lateral flexibility. They bend under a side force, displacing the tread and footprint sideways relative to the wheel rim. In a cornering situation, the tire therefore does not roll exactly where the wheel is pointed, but along a tangential course. This deviation of its rolling path is known as *slip angle*

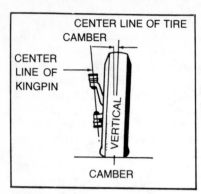

CENTER LINE OF TIRE

CAMBER

CENTER LINE OF KINGPIN

VERTICAL

CAMBER

Fig. 1-9. A front wheel with positive camber—it also has a positive scrub radius (distance at ground level between king-pin and tire center lines).

(though it would be technically more correct to call it a *creep angle*, since there is no actual sideslip, only a creeping further and further away from the aim of the wheel by a succession of tire footprints under the rolling action of the tire).

The slip angle increases more or less proportionately with the side force until it reaches a maximum—the point beyond which the tire loses its sidebite and goes into a skid. Depending on tire design and construction, this point can arrive suddenly or in a transitional fashion. The latter is, of course, preferable, as it gives the driver the earliest warning signals.

Side forces are not equal on the two front wheels of a car making a turn. Centrifugal force produces a phenomenon known as weight transfer, which can be described as an unloading of the inside wheel and loading of the outside one. Since side forces are proportional to the load, it follows that the tire on the outside front wheel must run at a greater slip angle than the tire on the inside one.

But with Ackermann steering, the inside wheel is aimed for a sharper turn and, with its reduced sidebite (due to the weight transfer), is being dragged along at an exaggerated slip angle. From this realization, an anti-Ackermann effect should be desirable. This applied only to situations of high side forces, however. In slight curves taken at low speed, some Ackermann effect is still desirable, though not actually needed (again, because of lateral flexibility in the tires).   Chassis engineers began to compromise by reducing the Ackermann effect and, in some cases, eliminating it.

For the same reasons, toe-in and camber in for deemphasis. As early as 1950, some European car makers specified zero camber and zero toe-in for certain models. The tire was counted on to compensate for minor alignment faults. Another factor was the public's demand for increased cornering speeds. And naturally, what's ideal wheel alignment for going down the turnpike is not the best for fast cornering. Weight transfer in a curve usually results in some amount of body roll. The car leans over towards the outside of the curve (in the opposite direction of motorcycles, boats and aircraft). This causes spring deflections that alter the toe-in and camber angles.

The main thing to avoid when cornering at high speed is positive camber on the outside front wheel—and rear wheel, too, for that matter. Wheels canted to the outside, on the car's outer side, cannot sustain very high slip angles before going into a skid. And to keep positive camber within limits in such situations, it may be necessary to align the wheels with a static negative camber setting.

Caster Angle for Directional Stability

How the different suspension systems affect wheel alignment when cornering will be discussed at some length in other chapters. Here we will continue our review of static wheel alignment. There are more angles to this, as my learned colleague Karl Ludvigsen would say.

First, there is the caster angle. Its use in the front wheels of the car is modeled on that of casters on a mobile piece of household furniture or the hand trolleys used for hauling luggage along railroad station platforms. The idea of the caster is a wheel whose spindle is offset from its vertical pivot. The offset assures that the wheel automatically aligns itself, like a rudder, with the direction of the push that moves whatever it is mounted under.

In cars, the caster angle assures a self-centering action in the steering system. This caster angle is the angular difference between the vertical line from the wheel hub center to the ground and the inclined line between the upper and lower mountings of the steering knuckle (ball joints). This latter line is the wheel's swivel axis, and its extension hits ground level some way forward of the tire contact patch center.

Like the furniture caster with its vertical swivel axis, the wheel with its inclined swivel axis automatically swings into a trailing position. Caster angle serves to bring the car's front wheels to a straight-ahead position and tends to keep them there. A greater caster angle will assure greater directional stability, but will cause heavier steering.

The upper and lower mountings of the steering knuckle (ball joints) also determine the steering-axis inclination (formerly known as king-pin inclination). It is the same as the swivel axis, but is viewed in a different plane. The swivel axis is seen from the side of the car (it runs in the fore and aft plane), while the steering-axis inclination runs laterally (and is viewed from the front of the car).

The inclined steering axis came into being because a vertical king-pin, placed some distance away from the center of the tire, would make the steering effort intolerably high. The steering axis would hit the road surface a long way outside the tire footprint, imposing a long radius of force action on the steering system. This radius is also known as *scrub radius*.

A king-pin placed vertically through the wheel hub so as to coincide with the tire's center line would reduce this radius to nothing. When the steering axis coincides with the tire footprint center, you have center-point steering. When the steering axis intersects ground level inboard of the footprint center, you have a positive scrub radius. Outboard, it's a negative scrub radius.

However, the usual design of hubs, brakes, and other parts makes it difficult, if not impractical, to have a king-pin in the middle of the wheel. Center-point steering can also be achieved by inclining the king-pin (or lining up the ball joints at an angle) so that the steering axis intersects the tire center line at ground level.

Most cars have a modest positive scrub radius which tends to assist the self-centering action of the caster angle. A negative scrub radius is used on some front-wheel-drive cars in the interest of assuring stability under braking (particularly with diagonally-split hydraulic brake lines).

The idea is to avoid any condition where a positive scrub radius could occur, even in the course of steering corrections during a braking skid. Any positive scrub radius does tend to throw severe stress loads into the steering linkage during braking; in front-heavy cars without drive to the rear wheels, a negative scrub radius on the front wheels can make all the difference between remaining in control of the car or losing it.

Details of the steering linkage and steering gear will be treated in a later chapter. In the next one, we will examine the main suspension links between the wheels and the car, namely the varities of arms, rods, bars and springs which serve to locate the wheels, and their influence on changes in wheel alignment and other criteria.

Chapter 2
What Links the Wheels to the Car

Wheels can be attached to the car in many ways. Modern suspension systems are made up of many components including control arms, springs, stabilizer bars and shock absorbers. Each component has a specific function to perform, but each is also only one element in a complete system. They must be regarded as players on a team where any action by one member affects all others.

The suspension system must be designed to permit each wheel to be moved up and down to the full extent required to keep all four tires in continuous contact with the roadway and, yet, restrict such movement to what is consistent with ride comfort requirements. The flexible element must be strong enough to support the load without using up wheel travel that's needed for bump deflections and, yet, soft enough to insulate the passenger compartment from suspension movements.

It is also desirable that the suspension system should allow the wheels to move back and forth to a small extent (horizontal compliance). Far more important, however, is the need for the suspension links to avoid tilting the wheels far out of their planes of motion (camber changes) during deflections. That's vital to the safe handling of the car at speed.

Precise guidance of the wheel throughout its full course of deflection should be assured to minimize not only camber changes, but also changes in caster, track and wheelbase. Driving, steering and braking put tremendous loads on the suspension members, and the links and their mountings must be designed with sufficient strength to handle these loads.

Essential Compromise of Functions

Because it is the wheels that propel, steer and stop the car, the suspension system must transmit the driving thrust and the decelerative forces of braking. Suspension members must be able to resist side forces acting on the car with minimal effect on the wheels and their attitude so that, during cornering, the wheels can remain consistent in their aim and the car's overall steering characteristics will not undergo any abrupt changes. Non-steered wheels must be suspended so that they are prevented from inducing steering phenomena into the car's behavior.

These multiple functions of the suspension system complicate things a great deal for the engineers who design and develop the car, for some of the requirements are in open conflict with others and even the best results contain some degree of compromise. And that's not all, for the suspension engineer has a lot of other things to consider besides geometry, loads and spring rates. He must take into account the production cost of the system.

Manufacturers of high-performance cars, whose customers have high demands in terms of cornering speeds and stability in cornering, accept higher costs than makers of family cars with low-speed capacity and no sporty pretenses. The reliability of the system must be assured, with minimal maintenance requirements and maximum life expectancy.

Next, the suspension engineer must consider the convenience of the installation. On a family car, maximum space must be allocated to the passenger compartment, whereas on a single-seater racing car, the designer has much more freedom (Fig. 2-1). It applies to all cars, however, that the suspension members must not get in the way of drive units or steering organs, or cause ground-clearance problems.

Finally, the suspension system must not limit the styling freedom of the body designers, steal trunk space or prevent practical station wagon arrangements from being made (Fig. 2-2).

The simplest way to attach a pair of wheels to a car, without forgetting flexible element, is to mount them at opposite ends of an axle and clamp the axle to the middle of two leaf springs with their ends tied or shackled to the car frame. This type of construction came into use on the first cars as an inheritance from horse-drawn vehicles where its drawbacks were less evident.

Axle suspension systems, with longitudinal semi-elliptic leaf springs, have many drawbacks whether for wheels that steer or wheels that drive.

Let's start with a look at a simple I-beam front axle.

The heart of the problem lies in the fact that the springs themselves are acting as locating members, and that introduces a big problem (see Fig. 2-3). Springs are intended to flex, and though their flexibility is needed in only one plane, it is the nature of leaf springs to flex also in other planes. They twist, for instance. And their up-and-down flexing is not accomplished without a fore-and-aft movement in the axle (and its wheels). They are ill-suited for taking up braking thrust, which tends to force the spring into an S-shaped profile, down at the front and up at the back, which usually results in negative caster with its attendant instability.

Long springs give a more progressive springing action and permit greater deflection that short springs, and that's conducive to better ride

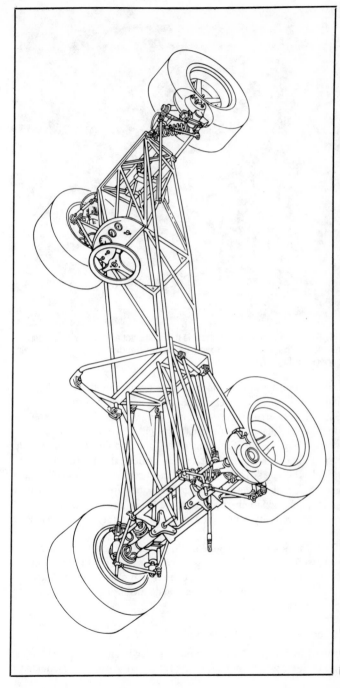

Fig. 2-1. Racing car designers have great freedom to choose suspension geometry without regard for space considerations, but are restrained by the need to save weight and avoid obstructing the airflow with springs and wheel-locating members.

Fig. 2-2. The modern sports-car chassis has compact and lightweight suspension members that do not encroach on space needed for the power train, passengers, or luggage.

comfort. But long springs are more likely to exaggerate their bending and twisting under braking, on very rough roads and when cornering.

Since the axle itself is carried as unsprung weight, it poses a problem for both ride and handling. Attempts to lighten the axle can result in weakening it to the point of inviting failure, and that's unacceptable. Axle breakage is almost guaranteed to cause an accident; it's, therefore, necessary to make the axle so robust that it will not break under any combination of loads and circumstances not only when new, but after many years of use. It must resist metal fatigue throughout the life of the car. And that means axles will of necessity remain heavy, aggravating the ride-and-handling problems.

The springs are also, to at least 90 percent of their mass, carried as unsprung weight. Multi-leaf springs have their centers of gravity in their geometrical middles, at or near the U-bolts that clamp them to the axle. Only towards the mountings at either end do the springs begin to become part of the sprung weight.

Spring Location and Weight Transfer

Since the front wheels need room to swing left and right for steering purposes, the springs cannot be attached close to the wheel hubs, but must

26

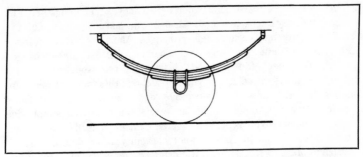

Fig. 2-3. A semi-elliptic leaf spring is capable of carrying out the double duty of locating the wheel and acting as a suspension spring.

be placed closer to the middle of the axle. That tends to give a very narrow spring base, which means that a small side force is enough to sway or tilt the body and chassis relative to the axle through a considerable roll angle, due to excessive weight transfer. That's uncomfortable for the vehicle occupants and can produce hazardous steering phenomena.

Fortunately the wheels do not undergo camber changes due to body roll with I-beam front-axle suspension. They remain upright and do not lose sidebite. But the axle is shifted from its static plane; it no longer runs perpendicularly across the car's longitudinal center line, but is pulled backwards on the outside as the spring is flattened and pulled forward on the inside as the spring curvature goes into a tighter arc. This puts an understeering effect into the car, but it can be easily overcome by merely turning the steering wheel to make a sharper turn, as long as the road

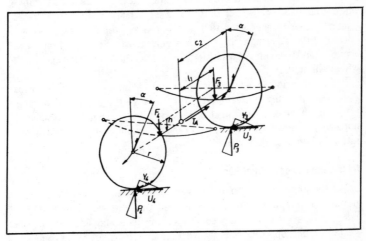

Fig. 2-4. A front axle carried on leaf springs transfers disturbances from one wheel to the other, and the springs provide flexibility in planes where rigidity is preferable.

27

surface is smooth, that is. For when it is bumpy, other things begin to happen, and they can happen on the straightaway just as well as on curves. That's because bumps cause camber changes. And since both wheels are attached to the same axle, single-wheel bumps will cause camber changes in both front wheels (See Fig. 2-4). If the right front wheel goes over a bump, the axle is raised at its right end, and that tilts the left wheel hub, putting the left front wheel at a positive camber angle for the duration of the deflection. Conversely, when one wheel runs through a pothole, the wheel on the other side is put at a negative camber angle. This will have a detrimental effect on the car's road-holding ability and handling precision if it happens on a curve, and can even affect the driver's control of the car on a straight road, as well as disturb ride comfort no matter where it occurs.

Soft Springs Lead to Wheel Shimmy

With nice, long, soft springs and heavy wheels, a series of bumps or ripples can set up a rhythmic cycle of axle movement relative to the frame, steering alternately left and right, setting up a strong shimmy in the steering wheel. Such shimmy can, in extreme cases, lead to loss of control.

This shimmy phenomenon is worth a little explanation. Soft springs reduce the suspension ride frequency, which means spacing out each spring deflection in time. This, in turn, reduces the axle tramp frequency and brings it closer to the natural frequency of the shimmy cycle.

When an axle begins to tramp, it sets up gyroscopic moments in both front wheels. These moments grow progressively stronger with each tramp because the wheel deflections are 180 degrees out of phase. And they affect the steering because the rising wheel toe-ins and the falling wheel toes-out. The wheels are forced farther and farther from the intended course, in rapid alteration, and this back-and-forth motion is transmitted to the steering linkage. The steering wheel is jerked out of the driver's hands, violently turning left and right in the alternating cycle of the wheel shimmy.

Stiffer springs will help reduce the risk of shimmy, but can also cause a worsening in the ride comfort, first, because the up-and-down wheel travel for a given shock load is lessened and, secondly, because the horizontal compliance (that is, the rearward displacement of the wheel on hitting a bump) is diminished.

Overlooked Longitudinal Loads

It is something that most drivers don't think about, but there is a horizontal component in the shock load that the wheel puts into the suspension system when it goes over a bump. The bump is an obstacle of a certain height, opposing the forward motion of the wheel. What comes first is a force that tends to push the wheel backwards relative to the rest of the car. It is overcome as the wheel rolls over the obstacle. No matter how soft the springs are, an extra force is required to lift the wheel assembly over it. This lifting force has a drag component, in the horizontal plane, which will be felt inside the car unless the suspension system offers adequate horizontal compliance.

Dynamic longitudinal forces (FA) are extremely important in reducing impact loads. Actual measurement of loads on test vehicles has shown that on poorly surfaced or gravel roads the total longitudinal loads can exceed the vertical loads into the ball-joints. The factors affecting longitudinal forces are:

- fore-and-aft suspension rate;
- longitudinal damping; and
- tire damping and deflection characteristics.

Leaf springs absorb this horizontal force by flattening out a bit, which stretches the distance from the forward spring anchorage point and the axle—not much, perhaps a quarter of an inch or half an inch at most. But as long as the suspension gives a little, that's enough to keep the horizontal component from playing havoc with the ride comfort.

There is also the risk of losing balance, handling precision and stability in braking, if there is too much horizontal compliance. The optimum compliance rate must be developed for each individual vehicle design, for too low a compliance rate can result in a vibration of the unsprung mass during deceleration, which feels very much like rough brakes. In addition, the shake level of the vehicle structure may be caused to deteriorate.

The high unsprung weight is perhaps the front axle's worst drawback and is the main reason it is no longer used on cars. On trucks and buses, it

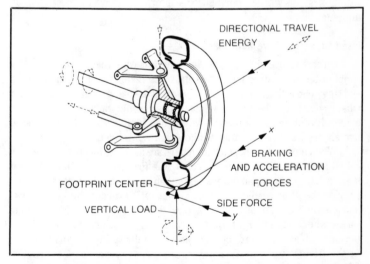

Fig. 2-5. With independent front suspension, the locating members are divorced from the springs. Smaller wheels reduce the distance between fore-and-aft forces acting through the tire footprint and those acting at wheel hub level.

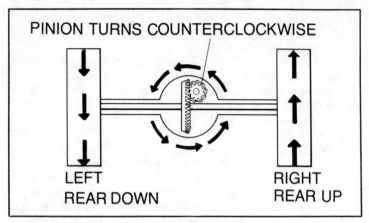

PINION TURNS COUNTERCLOCKWISE

LEFT
REAR DOWN

RIGHT
REAR UP

Fig. 2-6. Torque reactions are the worst problems of the live axle. Lift forces acting on the right wheel produce axle tramp on acceleration.

is still quite common. On heavy trucks, the axle weight does not lead to a disastrous ratio of sprung-to-unsprung weight, and the vehicle's great weight and load capacity make it necessary to use very stiff, multi-leaf springs which provide more positive axle location, as well as increased ride frequencies. Consequently, shimmy is not a threat, and other axle movements that influence the steering are easily countered on modern trucks equipped with power steering.

Cars have gone to independent front suspension (Fig. 2-5), and the various systems will be examined in Chapter 7. Now, we will take a look at the rear axle.

Disadvantages of Live Rear Axles

On cars where the rear wheels drive, it is called a *live axle*. It is not a simple I-beam construction but an elaborate casing that contains a pinion and ring gear with a differential, and two drive shafts, each connected to a wheel hub. It can be three or four times heavier than an I-beam front axle—all of it unsprung weight. If attached by longitudinal semi-elliptic leaf springs, they add further to the unsprung weight.

When no other suspension members are added to locate the axle, it is called *Hotchkiss drive* (from the car it was first used on). Here the leaf springs have far more ambitious tasks than on an I-beam front axle. They provide the same resilient support of the static load and the same lateral and longitudinal axle location relative to the chassis. But the forces acting on the wheels are different. Brake forces are lower since the front wheels perform at least two-thirds of the braking. But the springs also have to take up the driving thrust. In other words, they have to push the car to maintain steady speed as well as to accelerate.

The term live axle is particularly apt, for the presence of rotating gears and shafts inside it give it a life of its own, and their rotations sets up

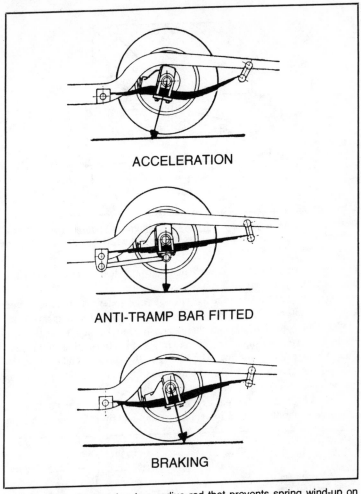

ACCELERATION

ANTI-TRAMP BAR FITTED

BRAKING

Fig. 2-7. An anti-tramp bar is a radius rod that prevents spring wind-up on accleration and resists brake torque effects during deceleration.

torque reactions in the axle casing which the springs are asked to resist (Fig. 2-6).

There are two kinds of torque reaction in a rear axle—the reaction of the axle housing to rotate backwards, in the opposite direction of the crown wheel rotation, and the reaction of the axle housing to spin like a giant propeller around its own center, rotating in the opposite direction of the pinion's rotation. The first leads to a lifting force in the differential nose-piece causing spring wind-up (Fig. 2-7), and the second leads to a lifting force on the right-hand wheel which manifests itself in axle tramp and attendant loss of traction (wheelspin). The springs are called upon to

31

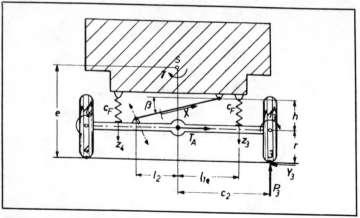

Fig. 2-8. A rear axle with coil springs needs a track bar for lateral location. Since it describes an arc, the axle moves left (relative to the car) in jounce and right on rebound (seen from behind).

provide the torque arm action needed to restrict axle rotation in two planes and are clearly not suited for such tasks.

As with the front axle, the leaf springs bend and twist so as to produce a steering effect. Though the springs can be mounted farther apart and closer to the wheel hubs, since there is no need for room to steer the rear wheels, there is an angular displacement of the axle relative to the body due to body roll in cornering. The flexing of the springs tend to steer the axle outwards—that is, in the direction away from the turn center, going off on a tangent, so to speak (See Fig. 2-8).

On a right turn, the rear axle tends to turn a few degrees to the left. The same happens with an I-beam front axle, and there it produces some

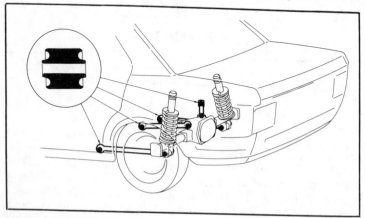

Fig. 2-9. Four-link rear suspension gives good control of axle movement. Rubber bushings are commonly used at all joints to absorb noise and vibration.

underster. But on the rear axle, it is oversteer. How come? Because steering the rear wheels left on a right turn tends to sharpen the turn. The tail end goes on a wider arc than the front, and that is an unstable condition.

Regardless of the shortcomings of Hotchkiss drive with regard to axle location, and its drawbacks in terms of unsprung weight, several current production cars persist in their use of this primitive type of suspension. More scientific forms of axle suspensions have been devised and are in common use on mass-produced cars in Europe and Japan, as well as in the U.S.A. (see Fig. 2-9).

But there is a general trend for the cars' rear ends to follow the fronts towards the use of independent suspension for the rear wheels, as well as the front ones. The individual systems will be dealt with after we have discussed the need for matching the front and rear suspension systems of a car to each other.

Chapter 3
Matching Front and Rear

No matter what type of car we are looking at, front and rear suspension systems should be matched to each other. It's common practice to have the front wheels do the steering and leave the driving to the rear wheels. With that type of layout, the duties of the front and rear suspension systems are broadly defined. But not just any kind of front suspension is a suitable partner for any kind of rear suspension. Each must be designed and developed with the other in mind, so as to create a harmonious whole: a well-balanced, well-behaved, stable and controllable car.

When front wheels both drive and steer, the same principles still apply. It remains equally true and valid that the front and rear suspension systems must meet the duties of their respective wheels and be matched for each other (see Fig. 3-1). This matching has to do with many things such as ride comfort, dynamic response, directional stability and handling safety. Those are things on which the car is judged as a whole, not on the strength of the individual virtues of the front and rear suspension systems. Front and rear suspensions, being parts of the same vehicle, interact to a very large extent. Examples of how they affect each other will recur frequently in the following discussion.

There are so many things going on concurrently in the suspension systems as the car travels along the road that it is even difficult to discuss one thing at a time. Ride and handling are inseparable as subjects of evaluation, since they depend on the same hardware.

At the risk of telling something less than the whole truth in each specific area, I will try to split them up. Ride will be dealt with at length in Chapter 5, entitled The Elastic Element, since it is a function dominated by spring stiffness or the lack of it.

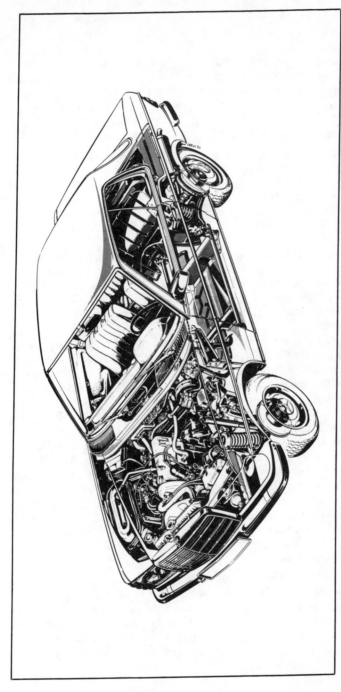

Fig. 3-1. Coil springs front and rear provide well-matched ride rates in the Mercedes-Benz 230 C, which has a forward weight bias unladen and a rear weight bias fully loaded.

While we are mainly concerned with handling in this chapter, it is inevitable that certain aspects of ride be brought into the discussion, however.

To get started, consider this: Unsprung masses have only one degree of freedom that counts—vertical wheel travel, which has a great deal to do with ride. The sprung mass, on the other hand, has three degrees of freedom: roll, pitch and yaw. These have to do with both ride and handling.

Pitch and Its Opposite, Bounce

If you think of up-and-down spring motions simply as bounce, then what is pitch? Both are determined by the interaction of the front and rear suspension systems. Pitch is a seesaw or rocking-chair motion, with the front end of the car bobbing up and down. In other words, the front springs alternate between jounce and rebound at a certain frequency, and the body oscillates around a horizontal axis that's located within the car's wheelbase, or at the rear suspension.

Some cars go into long series of pitch motions when traveling along a moderately rough road at a certain speed. When the front wheels hit a bump, the front end of the body lurches up. Then the rear wheels hit the bump, and the rear end of the car is kicked up. The body goes into pitch because it forms a bridge between its two separate suspension systems, front and rear.

Almost all cars undergo a pitch motion when crossing a large hump filling the whole width of the roadway, such as a raised level crossing with the railroad or a short-arched bridge. A car's tendency to pitch depends on its front/rear weight distribution (Fig. 3-2), spring rates and suspension geometry.

Bounce, on the other hand, may be thought of as the up-and-down motion of the total car in a level, non-pitching attitude, with the springs at all four wheels being deflected in unison. Actually, bounce is the opposite of pitch, occurring when the suspension system permits a cyclical oscillation of the body around an axis which may be located at the front suspension, or far ahead of the car, or at any point between. If the distance from the car to the bounce axis is great enough, front and rear springs undergo practically the same deflections at the same frequency. If the distance is short, the front suspension will participate to a lesser extent.

Side Forces Induce Roll

Roll is the sway or lean that occurs in the body due to side forces acting on the car. This side force is usually centrifugal force, generated because the car is yawing. Yaw is the same for a car as for a boat—a change of heading. Yaw is usually driver-induced (by steering wheel input), but can also be caused by roadway unevenness or a difference in tire-to-roadway friction between left and right wheels (especially under hard acceleration or braking).

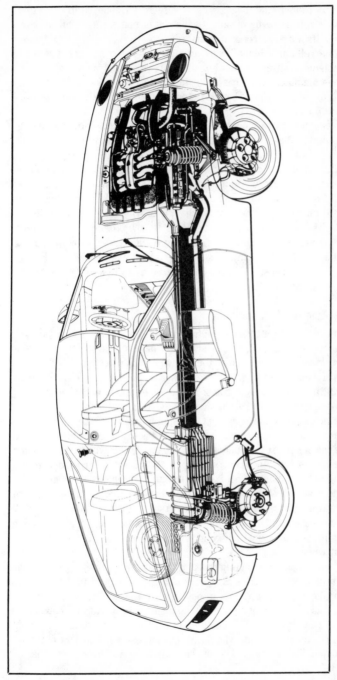

Fig. 3-2. With 50/50 front/rear weight distribution and passengers seated midway in the wheelbase, the Porsche 928 can use coil springs of similar stiffness front and rear.

Uncontrolled yaw means loss of directional control. In cornering situations, roll is a result of yaw. And roll, in turn, has a great influence on the car's stability and controllability.

Some roll can be helpful as a warning signal to the driver; but usually, the problems it causes for the wheels and tires make it desirable to restrict roll to a minimum. Most chassis engineers agree that for normal sedans and sports cars, roll should be restricted to 5 degrees for a lateral acceleration of 0.5 g.

In engineering parlance, side forces are measured as lateral acceleration in units of g—the symbol for gravity. It stems from the rate of acceleration of a free-falling object, identified as 1 g and corresponding to about 32 feet per second per second (32 ft./sec^2). A lateral acceleration of 1 g means that centrifugal force is equal to the weight of the car.

In everyday driving, few motorists go beyond 0.25 or 0.3 g. Then body roll and tire squeal warn them that the turn must be taken at a slower speed. Fast drivers in good sports cars may go up to 0.65 or 0.75 on the road. Grand Prix cars now are capable of taking curves with up to 2.2 g lateral acceleration. Racing cars corner without discernible roll. Family cars roll more than sports cars. Cars that sway (roll) a lot on curves are said to have poor roll stiffness.

Weight Transfer From Cornering

It is going to get a little complicated now, for in order to get the right idea about roll stiffness, it is best to first have a clear understanding of weight transfer in a car. Just as there is weight transfer towards the front under braking, and weight transfer towards the rear under acceleration (Fig. 3-3), there is weight transfer from side to side in cornering.

Weight transfer can occur simultaneously in both planes. On a curve, the principal direction of weight transfer is lateral. But some forward weight transfer also takes place because of the car's yawing attitude. The result can be almost total unloading of the inside rear wheel, light loading of the inside front, medium loading of the outside rear, and extreme loading of the outside front. It is important that the camber angle of the outside front wheel should be zero or negative under such conditions to give maximum sidebite. It is also important that the outside front spring should not use up all its travel in roll but have something left for deflection on bumps, and that jounce deflections should produce a minimum of toe-out in that wheel.

Centrifugal force, acting through the center of gravity, produces an overturning couple that is proportional to the force and to the center of gravity height. The force is proportional to the vehicle weight and increases as the square of the speed. To balance this overturning couple, the outside wheels must be given an extra load. In other words, the weight carried by the two outside wheels must be increased to equal the overturning couple. This extra load is conveniently borrowed from the two inside wheels.

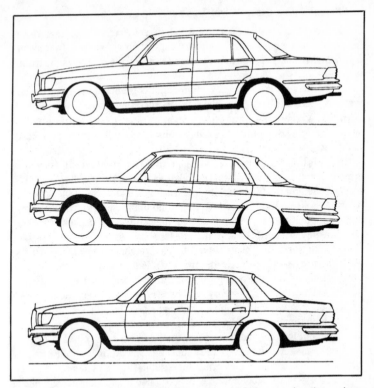

Fig. 3-3. Cars tend to nosedive on braking and squat during acceleration, unless suspension systems are designed with anti-dive and anti-squat geometry.

But weight transfer is not evenly shared between front and rear wheels. The exact division depends on body roll. Weight transfer is shared in the same proportion as the roll stiffness. If the front suspension has 50 percent higher roll stiffness than the rear suspension, the front tires will make a 50 percent greater effort to balance the roll couple. Roll stiffness distribution will be 60 percent in front and 40 percent in the rear, and the front wheels will get 60 percent of the total weight transfer.

Of course, the modern car has a body shell of great torsional rigidity, and it has to adopt the same roll angle front and rear, regardless of weight transfer and roll stiffness distribution. Roll angles are visually manifest and easy to measure and compare. But weight transfer has an inherent abstract quality since it is not apparent to the eye in a quantifiable amount. But it's there, and it is a determining factor in matching front and rear suspension systems. Fortunately, chassis engineers can calculate weight transfer on a firm basis of known elements under any combination of circumstances and plot the affect on a car's dynamic behavior of any change in suspension geometry and specifications.

Best Sidebite from Least Weight Transfer

Let us dwell a little longer on the question of weight transfer. Here's a simple rule to keep in mind: The front/rear distribution of weight transfer depends on the height and slope of the roll axis. That's not the whole truth, however.

As we have seen before, weight transfer is proportional to the center of gravity height (Fig. 3-4). It is also inversely proportional to the width of the track. A wide track means less weight transfer; a narrow track means more. Lowering the center of gravity means less weight transfer; raising it has the opposite effect.

Why is this weight transfer so important? You might as well ask why it is important for a tire to maintain its sidebite on a curve. We all agree that's a key element in road-holding. But the true explanation is just the opposite of what is commonly believed. It is minimum weight transfer, not maximum, that succeeds best in getting the car through the curve fastest.

The physics textbook says that the force necessary to move an object along a flat surface is proportional to the weight of the object and the coefficient of friction between the two surfaces. According to this theory, the tire's sidebite can be doubled by doubling the load it carries. That is not so in actual practice. The theory really does not apply to rolling tires. We are not dealing with pure rubbing friction, since the physical footprint is part of a continuous tread in motion.

The unit load (pounds per square inch of tire tread in contact with the road surface) comes into play, and it must be kept within limits. Extra load actually causes a loss of cornering force in the tire. A racing tire that can resist 1.45 g lateral acceleration with 500 pounds load, loses its sidebite at 1.33 g if the load is doubled. This was revealed by former Chaparral chief engineer Don Gates.

There are only two ways to increase the cornering speed (and resistance to lateral acceleration) of a car that is threatened with breakaway in the tires: reduce weight or increase the size of the tire footprints. Both *reduce* the unit load.

Roll Center Height Major Factor

How weight transfer affects the car's attitude depends on many things, and the first one to consider is *roll stiffness*. For a given side force, roll stiffness depends on three things. First and foremost is the *roll center height*. Second is the *spring stiffness*, and third the *stabilizer bar effect*.

To clear up the last item first, stabilizer bars are torsion bars that link the body to the suspension members on each side and resist roll (try to keep the body level). We shall look at some examples later.

Spring stiffness we have already talked about. The stiffer the spring, the more it resists any deflection, whether it is due to roll or any other cause.

Finally, there is the roll center height. What's a roll center? Look at a car going around a curve. It leans to the outside. This leaning is called

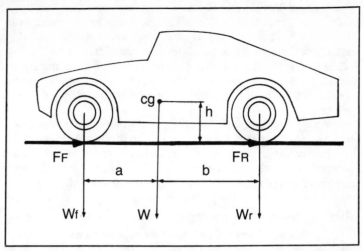

Fig. 3-4. Diagram for calculating forward weight transfer under braking shows that a low center of gravity and long wheelbase tend to reduce the pitch tendency.

roll (for it is a partial rotation). And any rotation takes place around an axis. A hammock has a roll axis that describes a straight line through the air between its suspension points (the knots on the ropes that hold the hammock to the trees.) Each knot, therefore, is a roll center. On a car, the front suspension system has its roll center, and the rear suspension has its own, too. The straight line drawn between them is the car's roll axis (see Fig. 3-5).

The higher the roll center, the greater the roll stiffness. In a car with its roll axis at ground level, all the roll stiffness comes from the springs and stabilizer bars. If the roll center at one end is high enough, roll stiffness may be so great that no stabilizer bar is required.

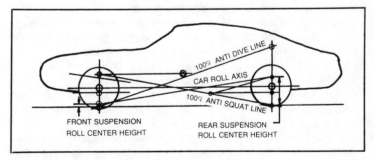

Fig. 3-5. Front and rear suspensions in combination determine a car's pitch axis as well as its roll axis. Tilt of front upper control arm assures anti-dive effect on the Oldsmobile Toronado.

41

Engineers can put the roll center almost anywhere they want it—at ground level, or at hub level, inbetween, and even below ground level. Where they want to put it depends mainly on weight distribution. Most cars have a sloping roll axis, low at the front and high at the rear. But on tail-heavy cars, better results are obtained with a high roll center in front and a low roll center at the rear. Examples of the reasons why, and the results, will be given in the chapters dealing with modern suspension designs.

When an engineer wants to increase roll stiffness, he has three ways to work: raise the roll center, increase the spring rates, and reinforce the stabilizer bars. He can use them individually or in any of the three design parameters will influence a host of others. The choice will depend on many and varied considerations, such as effect on ride comfort, steering response and handling stability. This brings us to the question of yaw.

Handling Precision From Yaw Angle

To make a car leave a straight path for a constant-radius curve, the car must be brought from zero yaw velocity to a steady-state yaw velocity and to a constant yaw angle relative to the intended path. The yaw velocity is introduced by the steering angle needed to make the car follow the curved path, and the yaw angle is required to bring the rear tires into a slip angle position where they can generate their fair share of cornering force (see Fig. 3-6). Yaw angle magnitude and the time needed to build it up and maintain it, are the factors which determine the response and handling pecision of the car.

It is important to realize that there is a time element involved in the development of weight transfer loadings. That's because the front wheels go into the curve first. The front tires assume a certain slip angle before the rear ones do. This gives rise to an immediate weight transfer at the front end. How the rear end reacts initially depends on the weight distribution in the car. In any case, the car's progress into the curve soon puts an outward slip angle into the rear tires, and weight transfer takes place also at the rear end.

Understeer and oversteer are transient responses in a car and can be defined as changes in vehicle yaw angles resulting from any change in side forces acting through the center of gravity. In practice, understeer means that the front tires run with wider slip angles than the rear ones. If the rear tires have the wider slip angle, the car will oversteer.

Because body roll induces changes in suspension geometry, most cars are subject to a certain *roll steer effect*. It exists in cars with I-beam front axles and live rear axles, as we have seen, and it exists also with independent front and rear suspension systems.

Roll steer is not always evil. It can be taken advantage of to improve the overall handling characteristics of a car. Concrete examples will be given in later chapters. Therefore, roll stiffness must not be too great to allow the body to lean to the extent necessary for generating the desired roll steer effect.

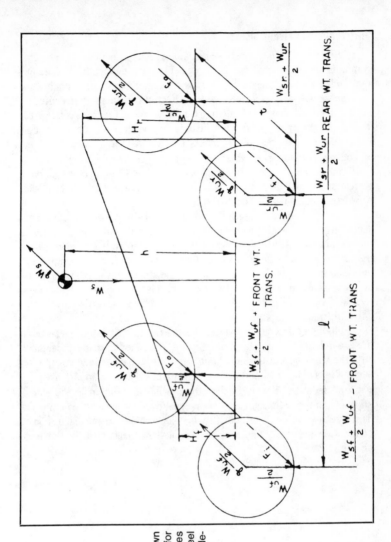

Fig. 3-6. Mathematical model drawn by Otto Winkelmann at Chrysler for calculating the cornering forces generated at each individual wheel takes into account 16 interacting design parameters.

43

Must Minimize Camber Changes

Regardless of what type of suspension system is used, it is important to minimize camber changes in the wheels during deflections caused by body roll. Camber changes in roll produce something that the engineers call *camber thrust*. It is a roll-steering effect of considerable magnitude, and it can work both ways. A positive camber angle of 5 degrees of a 165-13 front tire that carries a load of 500 pounds may produce a camber thrust of 75 pounds. That is a case of roll understeer. To counteract this, the driver will have to aim the front wheel one degree more sharply into the curve, which means cranking the steering wheel another 12 to 15 degrees (on a typical small car). A positive camber on the outside rear wheel would produce an equivalent amount of roll oversteer.

Camber changes during deflection are also to be avoided because they cause large gyroscopic precession torques about the steering axis. Such torques are picked up by the steering linkage and violently kicked back into the steering wheel. Not only is it unpleasant, but it can mask the important information that the wheels send back via the steering system and give the driver what he usually thinks of as road feel. It is a form of feedback, and its message stems from a quality in the rolling wheels that's called *self-aligning torque*.

Self-aligning torque is generated in the tire footprints and varies with the coefficient of friction between the tire tread and the road surface, and with the slip angle. It enables the driver to feel changes in slip angle, to sense a loss of friction, and to detect incipient breakaway in the tires. To get true feedback signals to the steering wheel, friction in the wheel supports, steering linkage and steering gear must be reduced to minimum. The steering system, consequently, is inseparable from any discussion of suspension systems. And that subject will be dealt with in the next chapter.

Chapter 4
Aiming the Wheels

How the driver aims the front wheels is an essential part of how he keeps the car under control. The mechanical connection is an intricate system that must work with high precision and give the proper change in aim for an input that balances the driver's muscular effort against how much he turns the steering wheel.

We know that the steering system consists of steering arms and a steering linkage, steering gear and a steering wheel (Fig. 4-1). As we have seen earlier, the arms and linkage are arranged to comply to some extent with the Ackermann theory. Now we shall discuss the total steering system and leave nothing unquestioned, right down to whether there isn't something that drivers will find better than a steering wheel for guiding the car (Fig. 4-2).

We could start with the steering wheel, but since we began our discussion of the whole car with a look at the road wheels, it is logical to go back to them once again for our examination of the steering system.

Apart from wheel alignment changes caused by suspension travel, the direction of each wheel is controlled by only one member: a steering arm. On most cars, the steering arm is attached to the front wheel hub by a keyway, locking taper and a nut. It is usually a separate forged-steel part. On some cars it is an integral part of a one-piece hub and steering knuckle. It extends backwards or forwards from the hub according to the location of the linkage, and its opposite end holds a ball joint or another type of joint.

These joints connect the steering arms to the left and right tie rods, whose other ends are attached to the relay rod. This relay rod runs transversely in the chassis and is usually a narrow-diameter steel tube, suspended between the tie rods and free to move left or right on command.

Fig. 4-1. A classic steering wheel with three spokes and plastic-covered rim is used on the Ferrari 365 GT, with horn on a lever, not in the wheel hub.

Fig. 4-2. Futuristic steering wheel and electronic instrumentation are major control-post elements of Bertone's Navajo on Alfa Romeo 33 chassis from 1976.

46

Steering Arm Length/Height Critical

The angle of the steering arms affects the relationship of the left- and right-wheel steering angles. The length of the steering arms plays a part in the overall steering ratio—the shorter they are, the less movement is needed to produce a given steering angle. Even more important is the height of the steering arm. It should not deviate from hub level.

If the steering arm ball joints do not coincide with the front-wheel center of rotation, the steering linkage will react.

If road surface unevenness produces wheel deflctions that cause the tie rods to move, a kickback type of steering wheel rotation will occur.

Commands to the relay rod are given, via a short drag link, by the pitman arm, whose shaft is part of the steering gear. The pitman arm shaft can be arranged to run vertically or horizontally, so that when it rotates, it turns the pitman arm through an arc, side to side if the shaft is vertical, fore-and-aft if the shaft is horizontal. The steering gear can be positioned to provide pitman arm movement in whatever plane is most suitable for the type of steering linkage the engineer chooses (Fig. 4-3).

The pitman arm and drag link pull the relay rod right or left. Its opposite end is attached to an idler arm, which is a blind copy of the pitman arm, pivoted on a bracket bolted to the frame. Its task is to assure accurate location of the relay rod relative to the vehicle frame. The relay rod carries

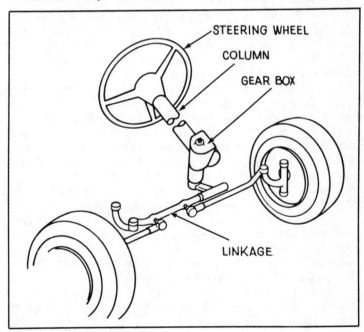

Fig. 4-3. Steering gear transforms rotation in the shaft to pull-push motion in the linkage.

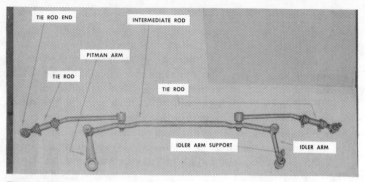

Fig. 4-4. Modern GM Saginaw parallelogram linkage is typical of current practice. Most GM cars now carry the steering linkage ahead of the front wheel axis (for stability).

two sockets, equidistant from its center, each of which is connected to equal-length tie rods (minor length adjustments are made by turning tapped sleeves around threaded sections of the tie rods at their outer ends).

Live Front Axle Steering Linkage

The linkage just described is known as the *parallelogram linkage*, (Fig. 4-4) and is used almost universally on cars with independent front suspension. On vehicles with I-beam or live front axles, a number of different and, usually, simplier linkages are in use. It may be useful in this context to take a quick look at steering linkages found on trucks, recreational vehicles and other special-purpose machinery.

In the fore-and-aft and cross steer linkage, the steering gear is usually located ahead of the front wheel axis. The left wheel carries a double steering arm—one part called a *bell-crank* reaching forwards to meet the drag link, and another pointing backwards at an angle, to connect with a relay rod running across the chassis to link up with a similar steering arm extending from the right front wheel hub (See Fig. 4-5).

Another type of fore/aft and cross steer linkage has been developed for use on cars with independent front suspension. The extra freedom of the individual wheels necessitated the use of an intermediate bell-crank pivoted on a fixed shaft attached to a frame bracket. Its two ends were attached to separate drag links, one connected in the normal manner to the left wheel bell-crank, and the other connected to the pitman arm. But this linkage has not found a widespread application.

Another type of linkage, named after its inventor, Haltenberger, is often used on vehicles having rigid front axles and is also adaptable to the Twin-I-Beam systems. (Ford uses it on light trucks). Here, the pitman arm is arranged to swing in the horizontal plane and works a single tie rod (to the right wheel). A separate socket on the tie rod, as near the center of

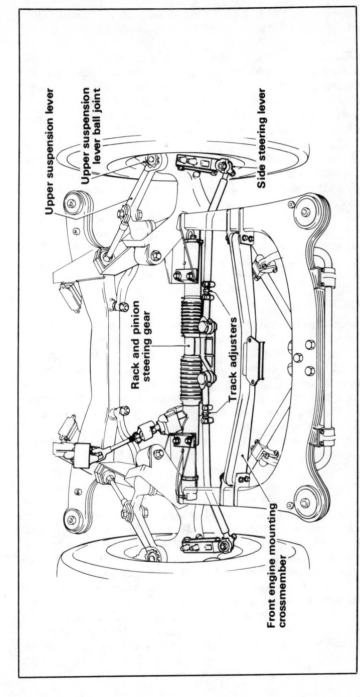

Upper suspension lever

Upper suspension lever ball joint

Side steering lever

Rack and pinion steering gear

Track adjusters

Front engine mounting crossmember

Fig. 4-5. Rolls-Royce steering system, seen from the front, has track adjusters as part of the drag links.

the chassis as possible, is connected to a second tie rod, running to the left wheel. Each tie rod is linked to steering arms extending from the wheel hubs.

The Haltenberger linkage can also be arranged with a vertical-plane pitman arm (See Fig. 4-5). In this design, the pitman arm pulls a drag link whose opposite end carries a ball joint. From this ball joint, a bell-crank, pivoted in a mounting on the frame, swings the end of a tie rod that carries a socket for connecting the second tie rod. The tie rod ends carry ball joints that hold the steering arms in the usual way.

No tie rod linkage can provide perfect Ackermann geometry, since it meets the design goals only in certain positions—straight ahead, one or more points on a left turn and corresponding points on a right turn. In intermediate positions, the wheels only momentarily assume right angles to the actual turn radius. For instance, when the inner wheel is steered 30 degrees to one side on a car with a given wheelbase, the outer wheel is steered 26 degrees to the same side, but should only have a 24-degree steering angle. There is a certain deviation between the locus of Ackermann steering and the locus of tie-rod geometry known as *tie-rod-geometry error*. (See Fig. 4-10) It is small enough that the tires are usually able to absorb it.

Novel Rey Linkage

Another type of linkage exists which provides true Ackermann geometry throughout the full range of steering angles (and, incidentally, allows the inside front wheel to be turned a full 90 degrees). (See Figs. 4-11 and 4-12). It is the invention of Andre Rey of Albi, France, who holds French, British and U.S. patents for his linkage. It has been examined by a number of car and truck builders including Renault, Fiat, Volkswagen, Ford, Peugeot, Citroen and Mercedes-Benz. But none have found the advantages to be worth making the change.

Since we are concerned with principles as well as practice, it is relevant to look at Rey's invention and the theory behind it.

The front wheels have normal steering arms, each attached to a long tie rod. Both tie rods are attached to the same ball joint, which is mounted on a pinion. The pinion runs along a rack with angled end sections, so that both tie rods are pulled back towards extreme-steering angles. In addition, the whole rack assembly is mounted so that it is allowed to tilt slightly left or right and, thereby, provide variable length for the tie rods in a pattern that forms an isosceles triangle between them and the straight line between the steering-arm ball joints at all times. (See Fig. 4-13).

The existence of such a triangle (isosceles means having two equal sides) is the key to true Ackermann geometry and avoidance of the tie-rod error. The mathematical formulas that prove the truth of Rey's theories are undisputed. It is certain that the geometry of Rey's steering assures full conformity with the Ackermann theory, for the axes of the front wheels invariably coincide with the rear wheel axis all the way from the straight-ahead position to the maximum steering angle.

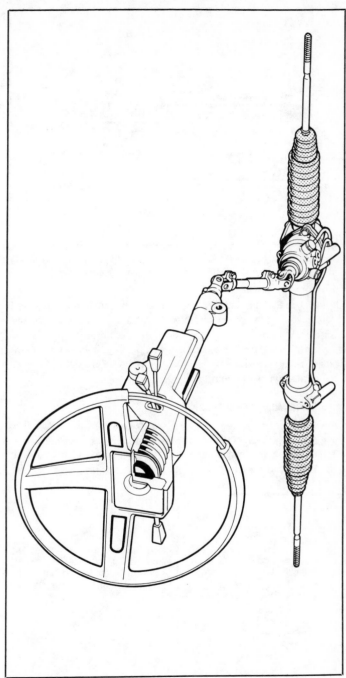

Fig. 4-6. Articulated steering shaft on Alfa Romeo 6 is combined with rack-and-pinion steering gear located behind the front wheel axis.

51

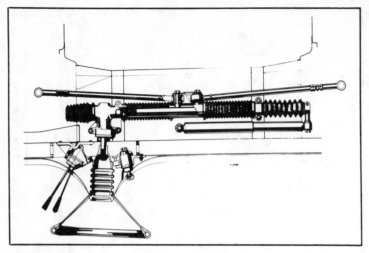

Fig. 4-7. On the Lamborghini P-250 Urraco, the rack-and-pinion steering gear is located in the cowl structure, with long drag links and steering damper.

It is certain that the Rey steering system, as patented, poses installation problems. On front-wheel-drive cars it will almost certainly cause the steering arms to be moved to a higher level (to avoid interference between the drive shafts and the steering linkage) and, thereby, invite a higher risk of kickback in the steering wheel. Finally, costs are certain to be higher than for normal types of steering gear and linkage.

Steering Gears Multiply and Filter

A variety of steering-gear types are in common use today. The steering gear is the heart of the steering system and it has a dual function:

1. To multiply the torque input from the steering wheel and transmit it to the linkage without lost motion in a constant or progressive ratio;

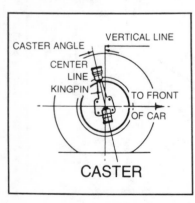

Fig. 4-8. Caster angle is obtained by tilting the king-pin (steering swivel) backwards at the top, so that the tire footprint center trails the steering axis.

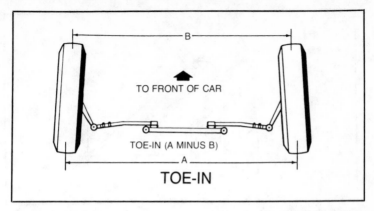

TO FRONT OF CAR

TOE-IN (A MINUS B)

TOE-IN

Fig. 4-9. Toe-in is used to counteract the force which tends to make the wheels toe out at speed.

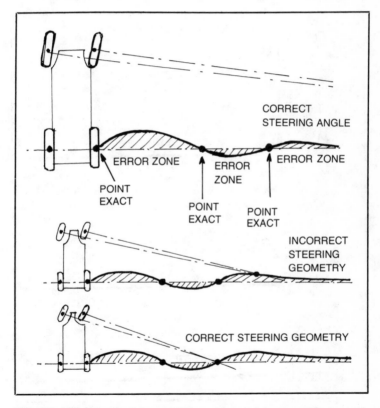

CORRECT STEERING ANGLE

ERROR ZONE

ERROR ZONE

ERROR ZONE

POINT EXACT

POINT EXACT

POINT EXACT

INCORRECT STEERING GEOMETRY

CORRECT STEERING GEOMETRY

Fig. 4-10. Actual steering linkages fail to provide true Ackermann effect except at certain points along the full range of steering angles, with the result that the wheels go from one error zone to another as the turn radius changes.

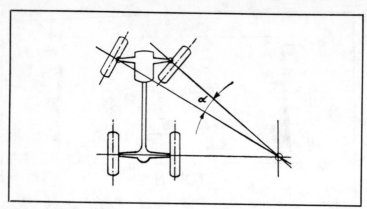

Fig. 4-11. Oversimplified Ackermann/Jeantaud diagram shows the basic idea of having a common turn center for all wheels, including the rear ones.

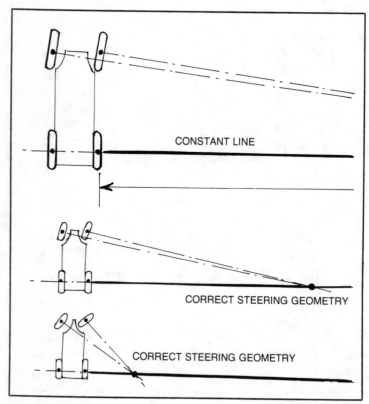

CONSTANT LINE

CORRECT STEERING GEOMETRY

CORRECT STEERING GEOMETRY

Fig. 4-12. With Rey's steering linkage, the wheels are continuously aimed with perfect Ackermann geometry. Rey's linkage also permits 90-degree steering angles.

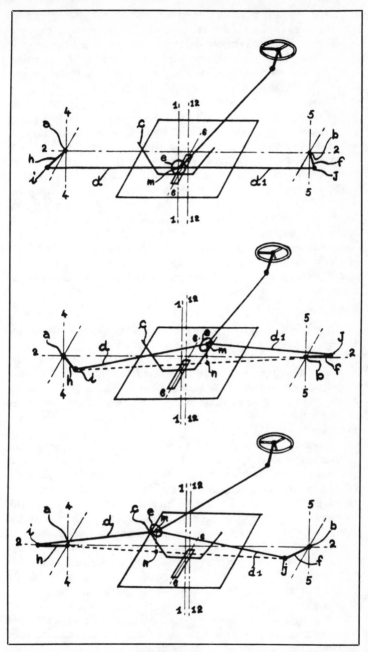

Fig. 4-13. The precision of Rey's steering geometry relies on a moveable rack so as to vary the effective length of the drag links.

55

2. To filter out road shock but permit feedback of self-aligning force to tell the driver how much of the car's cornering power he is making use of.

The steering gear is mounted on the vehicle frame, front sub-frame, or a mounting bracket in cars with unit-body construction. Its mounting must be strong enough to withstand the reaction forces from the steering linkage, as well as the input from the steering shaft.

When roads were rougher than today, it was very important that the reversing forces from bumps in the road should not be kicked back to the steering wheel—that could cause loss of control. To minimize this problem, most engineers used worm-gear types of steering gear. It is a peculiarity of worm gears that they run with more friction in reverse than forward, and if the helix angle on the worm is small enough, a worm gear will be irreversible. Use of a worm gear would enable a driver to resist the reverse forces from bumps and hold the steering wheel at any angle, regardless of road surface unevenness, while getting the benefit of high torque multiplication in the turn direction.

As roads got better and speeds higher, it became more important to provide the driver with self-centering action and feedback information from the wheels about the self-aligning torque in the tires, than to keep reversing forces out of the steering wheel.

Five Types of Steering Gear

Nevertheless, worm-type steering gears are still in common use. One of the surviving types is called *worm and roller*.

Worm and roller steering gear consists of an hourglass worm on the steering shaft engaged with a rotating toothed follower on the sector shaft. The pitman arm converts the rotation of the sector shaft into a force acting on the drag link that forms part of the steering linkage. The pitman arm is splined to the sector shaft and connected to the drag link with a ball-stud mounting.

Another type of steering gear, more common on commercial vehicles than on passenger cars, is called *cam and lever*. Cam and lever steering gears use a constant-diameter worm, which somehow has come to be called a cam instead, mounted on the steering shaft, meshing with one or two conical studs mounted on a lever that pivots on the sector shaft. With two studs, the system is sometimes called *cam and twin lever*. Stud motion along the cam produces rotation in the lever. This rotation is repeated in a 1:1 ratio in the pitman arm (which is usually twice as long as the lever).

Large cars (and especially cars with power steering) usually have *recirculating-ball* steering (Fig. 4-14) which is notable for having the lowest friction of all systems known. In this type of steering gear, the lowest section of the steering shaft takes the form of a worm gear, with balls running in the helix grooves formed there. The balls are held in their races by a sleeve with its inner surface made up with matching helix grooves.

This ball nut moves up or down the shaft when it is turned left or right. The outer surface of the ball nut carries a rack which meshes with a sector

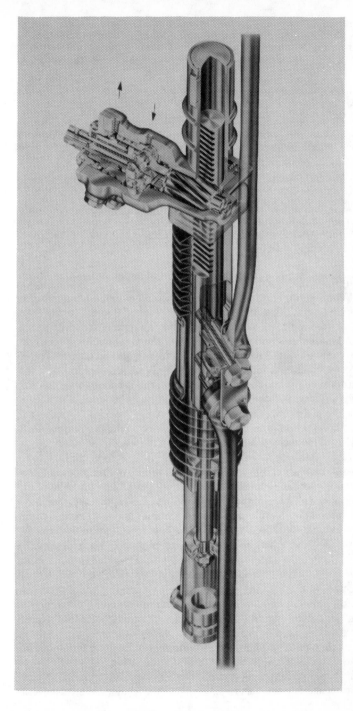

Fig. 4-14. Recirculating-ball power-steering gear by ZF is light and compact. Similar systems are used by BMW, Alfa Romeo, Peugeot and Volvo.

57

gear on the sector shaft. The linear motion of the ball nut is transformed into rotating motion of the sector shaft, which again turns the pitman arm.

One popular type of steering gear has no pitman arm. This is called *rack and pinion* and is a system that has long been popular in sports cars and has spread to small cars of all types (and some larger ones). (See Fig. 4-15). The steering shaft carries a pinion with helical teeth at its bottom end. This pinion engages with a rack that is mounted in a rail running across the chassis. Rotation of the pinion produces side-to-side motion in the rack. The rack serves as a relay rod, with its ends carrying ball joints connected directly to the tie rods.

Rack and pinion steering is chosen by many chassis engineers for its direct action and because it permits simplification of the steering linkage. In the past, many auto makers shied away from it because of its high reversibility, but today this is not a problem. Helix angles can be set to avoid road shock transfer and still allow sufficient feedback for the driver to have proper "road feel."

Steering Wheel and Shaft Mounting

The steering shaft varies in length according to the location of the steering gear and its distance from the driver's seat. Recent GM cars have had the steering gear mounted ahead of the front-wheel axis, with a long and articulated steering shaft. At the other end of the scale, take the Fiat 1500 roadster of 1963 which carried the steering gear on the front end of the cowl structure, with a vertical and transverse-action pitman arm and long drag links and tie rods to each steering arm. Some cars with rack and pinion steering also have articulated steering shafts because the engineers want the pinion located as near the center line as possible, while the steering wheel must be on the left, and using a straight, integral shaft would put too much of an oblique angle on the wheel.

It is taken for granted that a lateral twist in the steering wheel is to be avoided, but many cars have been produced with up to 10 degrees variation from a straight-ahead mounting. As for the flat V8 upright angle of the steering wheel, it was formerly determined by the length—and, therefore, the slope-of the steering shaft. Today, the shaft can be articulated to put the wheel at any chosen angle. What is best? We'll get to that shortly. First, a few more facts about the mechanical aspect of its mounting.

The steering wheel has a metal hub which is splined to the steering shaft. From this hub extends a metal armature that forms the load-carrying structure of the steering wheel. It is enveloped in plastic, rubber or another material of suitable temperature and chemical characteristics. The wheel hub on some cars also contains mechanisms for turn signals, horn, wipers, hazard flashers and rear-window defrosters.

Below the wheel hub, the steering column carries a lock assembly which is combined with the ignition system and transmission selector on some cars with automatic transmission. On cars with a manual column-shift, the mechanism is also combined with the ignition and steering-shaft

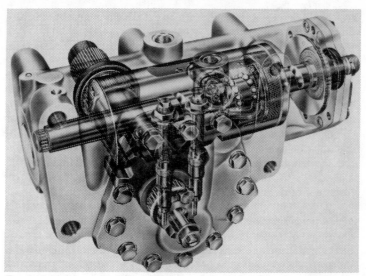

Fig. 4-15. High-precision ZF rack-and-pinion steering gear for sports cars works with a very slight helix angle. Power assist is built into the pinion shaft.

lock. Farther down on the steering column is a collapsible section that prevents intrusion of the top of the shaft (and the steering wheel hub) beyond a certain limit into the passenger compartment.

How big should the steering wheel be? A large diameter is required for the steering wheel to reduce the torque required for a given input on the steering shaft. On the other hand, a small diameter is needed to provide clearance for the driver (knees, lap and chest) while being within comfortable reach of his hands. With power steering, the wheel can be made as small as clearance needs dictate, without running into manual-effort and torque problems. Sixteen-inch steering wheel diameter was normal for American cars 25 years ago; today, 10-inch and 12-inch steering wheels are common.

Once it was thought that deep-dishing of the steering wheel, with the rim raised from the hub, was a safety factor (in protecting the driver's chest). Now, it has been established that a relatively flat wheel profile, with a large-size padded hub, gives the best protection.

Two-spoke steering wheels are the most common today. They permit strong support with little blocking of instruments from the driver's view. Some produce three-spoke wheels, but the four-spoke wheel has just about disappeared. One make, Citroen, has found its own solution in the single-spoke wheel, having produced a variety of such designs since 1955. The latest one has a wide, padded spoke extending back and down from the hub.

Telescoping and Tilting Features

A steering shaft that telescopes to provide a choice of reach for the driver is perhaps more important than the angle of the steering wheel. A system of telescoping tubes that can be extended or compressed, with a clamp-type locking device, has been available since the 1930's.

In the mid-Sixties, American manufacturers began offering steering wheels that offered both a variable angle and variable reach—*Tilt-and-Telescope wheels*. The telescoping feature was handled very well, but the tilt arrangement had the pivot point in the wrong place. It was positioned just below the steering wheel hub (Fig. 4-16) so that the wheel swung upward for a flat bus-type angle, when the lowest position would seem proper from an orthopedic standpoint. Similarly, its most upright angle could only be obtained at the lowest point of vertical adjustment, because the wheel swung downwards instead of upwards when tilted towards the modern race-car position. The answer is to move the pivot point to the lower end of the steering shaft. (See Fig. 4-17).

The first car in recent history to offer this type of adjustment was the 1965 Triumph 1300. It was a small front-wheel-drive family sedan with a steering column that provided two inches of height adjustment and four inches of axial travel. The vertical adjustment was restricted to two inches because the pivot point at the lower end of the steering column was a crude universal joint—not the constant-velocity type.

Similar arrangements were standard on the 1970 Citroen SM, with a vertical adjustment range of 2-1/2 inches and a four-inch telescoping range; the 1972 Alfasud (without telescoping), 1974 Citroen CX; and 1975 Alfetta. This type of design is compatible with federal requirements for collapsibility (rearward displacement of the steering wheel in a collision impact). It might even offer advantages in the safety area, depending on positioning of the pivot point and articulation of the steering column.

Fig. 4-16. Tilt-wheel articulated near the wheel hub forces undesirable and unnatural attitude changes on drivers. Wheel with flattest angle is too high, and wheel with steepest angle is too low for comfort.

Fig. 4-17. Steering shaft with low pivot point gives helpful height adjustment for steering wheel without significant changes in wheel attitude.

Alternatives to Steering Wheel

Now, perhaps we must ask whether we should seek to replace the steering wheel. It can cause injury in an accident. Some drivers find it interferes with comfortable seating and some designs partially block the driver's view forward and of the instruments.

What else could one use? There's always the tiller, as used in most pre-1900 cars, but it has even worse drawbacks than the steering wheel.

The airplane-type joystick, then? GM tried it first on a 1957 Chevrolet and made it a feature of the experimental Firebird III gas-turbine car of 1958. It was unsafe, in the sense that the driver had to be prevented from making excessive inputs at high speed; and, since that was done by a variable ratio which gave less response as the car went faster, it lost all chances of giving rapid countersteering in a skid. And in case of power-assist failure, the car was unsteerable. New versions were tried by Pontiac in 1965 before GM gave it up altogether.

Wrist-twist steering was used experimentally by Mercury in 1965. The steering column ended at the base of the instrument panel, where a yoke carried two extensions with chains and sprockets, each carrying a small wheel, about five inches in diameter, with their pivot points not in

61

the center but on the rim. It could not work without power steering, and a backup system was provided. But it was never developed for production.

Saginaw Division of GM made a dial-steering system for Oldsmobile in 1962, with a simple dial mounted on a pod rising from the center console. It led to nothing. In 1967 Buick and Saginaw were testing their own version of wrist-twist steering. Chrysler made a dual handgrip version of wrist-twist steering for its 300 X of 1966, and it was quickly abandoned. Few drivers can easily twist their wrists more than 90 degrees each way, and that posed power-assist problems in addition to needing a backup system.

It really looks like we're stuck with the steering wheel, and perhaps we should be glad. It probably is the best device for the job. I am supported in this belief by Bob Love, a top-ranking Chrysler engineer who spent years as head of the steering lab. He told me about one of his experiences from World War II when he was engaged on design work for the Chrysler tank plant, specifically in connection with gun turrets. The objective was to find the quickest and yet most accurate way to turn the turret to train the gun on its target. Bob Love went ahead and tried all known mechanisms, from levers and pull-rods to cranks and back again. The result of all this experimentation was the adoption of a steering wheel. It gave the best combination of precision and rapidity—by a wide margin over all other devices tested.

Time Best Basis to Rate Steering Systems

It is difficult to come up with an overall rating for a car's steering system. What we have to go on is mostly design specifications, such as the steering gear ratio, or the number of turns of the steering wheel needed to go from full-left lock to full-right lock, or vice versa.

The steering gear ratio, in itself, is almost meaningless. It simply indicates the gear reduction from steering shaft rotation to pitman-arm rotation. The overall steering ratio is a more meaningful number, for it takes into account steering-arm length and pitman-arm length, and relates steering wheel input directly to the steering angle of the wheels.

Lock-to-lock is not a very good criterion for measuring steering action, any more than is simply quoting the steering-gear ratio. Both ignore the maximum steering angle and, therefore, the turning diameter. All these factors are related. Even the wheelbase plays a part, for at a given steering angle, the car with the shorter wheelbase will make a tighter turning circle than a car with longer wheelbase.

In the final analysis, what really counts is the subjective impression gained by the driver as to whether the car has "good" steering or "bad" steering. Some drivers will talk about "fast response" and "slow response" when they really mean simply "fast-geared" or "slow-geared" steering. Steering response is not a measure of the steering system alone, for it involves various aspects of handling (yaw velocity, for instance) and is, therefore, dependent on factors beyond the steering system. Steering response is actually measured not in degrees, ratios, or distance, but in *time*—it is the time the car needs to respond to a certain steering wheel input at a given vehicle speed.

Of course, the overall steering ratio plays a part in determining the steering response. It is very difficult to obtain short response time with slow-geared steering, for instance. Since fast response is a vital factor in the active safety of the car (accident avoidance, such as in skid correction, steering around an obstacle, etc.) most engineers aim for relatively fast-geared steering. How fast-geared it is practical to make the steering system on cars without power steering has to be a compromise with the maximum tolerable muscle effort needed to turn the wheel.

We all know that it is very hard to turn the steering wheel when the car is standing still. As soon as the wheels are rolling, it becomes much easier. But it is also important that the effort should remain fairly constant throughout the car's speed range. Steering that gets progressively looser as the car gains speed is a hazard to the driver's control situation.

Loose steering at speed can usually be corrected by adjusting front wheel alignment, or taking up slack, play or lost motion in the steering gear and linkage. Of all existing types of steering gear, the rack and pinion is the least prone to looseness. Slow-geared units are more likely to develop slack than fast-geared ones.

An overall steering ratio of 15:1 is fast; 25:1 is slow. Since the linkage almost invariably adds something to the overall ratio, a steering gear of about 12:1 will be needed for an overall ratio of 15:1. Correspondingly, the car with a 25:1 overall steering ratio will have a steering gear ratio of about 21:1.

A sports car with a 15:1 overall steering ratio will give no more than three turns lock-to-lock with a wheelbase of 104 to 108 inches and a turn diameter of 30 to 33 feet. A heavy sedan with a 25:1 steering ratio, built on a wheelbase of about 120 inches and having a 40 to 44-foot turn diameter, will end up needing 4-1/2 to 5-1/2 turns lock-to-lock.

The average driver cannot tolerate faster-geared steering on a car of that type, since it would demand too great an effort and become very tiring to drive. With power steering, the ratio can be chosen independently of such considerations, and large station wagons have been produced (notably by Pontiac and Buick) with steering needing no more than 2.9 turns lock-to-lock (with Saginaw power steering).

Power Steering Permits Variables

But fast-geared power steering introduces a new risk. An inattentive driver may inadvertently crank the steering wheel through a wide angle, lose control and cause an accident. To reduce this risk, Saginaw and the GM car divisions developed variable-ratio steering. Here the steering gear gave a slow ratio (say, 18.5:1) in the middle, dropping to a very fast ratio (say 12.5) towards the extremities. This combination would give slow response for easy turns, and make easy work when parking or maneuvering in tight places.

Various transition curves were tried to establish the ideal point of changing to a faster ratio and the rate of progressivity. No absolute conclusions could be reached, but it is certain that variable-ratio power steering does not give adequate protection against loss of control while simultaneously introducing a new risk of overshoot in deliberate turns.

A different and far more logical approach has been taken by Citroen. Instead of a variable ratio, it is the power assist that is variable. It was first used on the 1970 SM and later adopted for the CX. The SM was very fast-geared, at only two turns lock-to-lock. The CX needs 2-1/2 turns.

The hydraulic power assist is arranged to provide full boost at standstill or low speed, for easy parking and maneuvering. As speed is increased, the amount of boost is gradually diminished, so that it becomes physically impossible, even for the best-muscled driver, to make a sudden turn at high speed. The power-steering unit also provides a self-centering assist, so that on long curves the driver has to actively resist the return to a straight course. When the car is parked and the engine turned off, the front wheels automatically return to the straight-ahead position.

Does it work in skid recovery? These front-wheel-drive cars rarely break traction or sidebite, but can be tricked by a sudden change in road surface (from asphalt to gravel, for instance) in a curve. I have found the word of Pierre Dupasquier, chief of Michelin's racing-tire program, that the Citroen power steering is fully capable of giving the desired action without pump catch.

Pump catch is something that used to occur in American cars with power steering when wide-oval tires came into use in the mid-Sixties. On quick steering reversals, the power steering pumps sometimes failed to provide the necessary boost and the driver would feel the steering wheel momentarily locked in his hands. Variable-ratio power steering aggravated the frequency of pump catch until new systems were developed. Citroen never had a problem, for its power steering gets it boost at high, constant pressure from a central hydraulic system with an ample reservoir (and not simply a small hydraulic pump driven by a V-belt).

There has been a disturbing trend in the 1970's to install power steering on smaller cars in order to cover up for flaws in the chassis engineering. Subcompact cars with four-cylinder engines do not have nearly the weight problems of large cars from the era before power steering became available, and a car like the 1950 Cadillac was quite manageable without power steering. When we see little Mustang and Monza sport coupes of today equipped with power steering, it's a cop-out. Instead of fixing up the basic problems wherever they are in the suspension geometry, front wheel alignment, or steering gear and linkage, the engineers have masked their existence by bolting on a power-steering package.

Power steering should be avoided if at all possible, for even the best of systems—including Citroen's, the ZF (BMW and Volvo), Saginaw and Mercedes-Benz—cause a certain loss of "road feel" which is so precious to the driver.

Chapter 5
The Elastic Element

The modern car has several stages of elastic elements interposed between the road surface and the passengers. Tires flex and, thereby, provide some measure of elasticity. Occupants ride in seats that have flexible support. But the major elastic elements are the *suspension springs* (Fig. 5-1).

Springs are necessary, as we have seen, to keep all wheels on the ground at all times. If the springs are too stiff, one or more wheels can still easily lose contact with the road surface. Soft springs, therefore, are a basic necessity for proper traction and sidebite.

Soft springs are also necessary to assure the minimum reaction in the car structure when the wheels run on a bumpy surface, so as to minimize variations in the forces exerted by the springs against the frame or body shell. On the other hand, springs must be stiff enough to prevent frequent bottoming which would damage the bump stops as well as cause acute ride discomfort. Low spring rates also lead to more body roll and a loss of stability in cornering and a greater change in vehicle ride level and attitude between unladen and full-load conditions.

The matter of selecting the right springs for a car would be an easy matter if the springs did not also have to support the load of the full sprung weight plus the payload. But there is no other way to install suspension springs and have them perform the basic function of insulating the vehicle structure from the motions of the road wheels.

Small Car Springing

Spring selection for small cars is more difficult than for big cars, because the payload represents a higher proportion of the total load. Proportional variations in load are greater than with large cars. This is

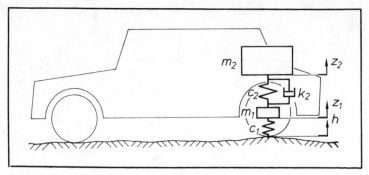

Fig. 5-1. Suspension springs are not the only elastic element, since tires also flex and provide an initial absorption of obstacles. But it's the suspension springs that provide the full extent of wheel travel.

called a *high load ratio*. A car with a curb weight of 2,500 pounds made to transport five people with an average weight of 150 pounds plus 400 pounds of luggage, has a load ratio of 1.46:1 compared with 1.34:1 for a car with a curb weight of 4,000 pounds built to carry six passengers and 450 pounds of luggage.

With a high load ratio, the springs have to provide both greater flexibility and greater resistance than in cars with low-load ratios. The high load ratio makes a greater demand for softness in the spring when the driver is alone in the car and also a proportionally greater demand for stiffness when running with a full load.

Light cars also pose a greater problem for the engineers in terms of keeping down the unsprung weight. The ratio of sprung-to-unsprung weight is another key to defining the working conditions of the springs. With a 10:1 ratio of sprung-to-unsprung weight, the vertical force transmitted to the sprung mass is only one-tenth of the force acting on the unsprung mass.

If a small car is built with the same type of rear axle and independent front suspension as a typical large car, there will not be a big difference in unsprung weight between the two cars. Say it's 325 pounds in the smaller car and 375 pounds in the big one, but the small one has a sprung mass of 2,175 pounds unladen, while for the big one it is 3,625 pounds. The sprung-to-unsprung weight ratio for the big car is 9.67:1, while in the small one it is 6.69:1.

That's why the engineers on a small-car project must use more imagination and accept higher relative cost in the design of the suspension systems in order to provide big-car ride comfort while preserving small-car roadholding.

Springs Deflect Energy

Springs—they come in different types, shapes and materials. Broadly they can be separated into two categories: metallic and non-metallic. In this chapter we will restrict our discussion to metal springs because other springing media are usually part of suspension systems that are conceptually different, often incorporating automatic level control, and must be treated by a systems approach. Metal springs, which are used in about 95 percent of 1980-model production cars made in the Western world, can be regarded as separate components with easily quantifiable characteristics.

What is a spring? It is a mechanical device that deflects energy by changing shape in proportion to the force that is acting on it. It has an ability to deflect load at a predetermined rate and recover its original shape when unloaded. As used in clockwork, the spring stores energy; as used in auto suspension systems, it expends and transfers energy.

Leaf Springs

Metal springs come in three basic types: *leaf springs, coil springs* and *torsion bars* (Fig. 5-2). Leaf springs usually consist of several lengths of steel strip cut from the same roll, assembled by lamination, joined by a bolt through their geometrical center and clamped together. Only the main leaf has eyes for mounting pins or shackles. The extremities of the supporting leaves are usually square-cut.

Each leaf has its own rate of flexing, depending on length, width and thickness. United in a multi-leaf assembly, they provide progressively stiffer action with increased deflection (Fig. 5-3). Tapered leaves provide some individual progressivity, due to a gradual increase in thickness towards the center. Single-leaf tapered springs have been used for some cars, and for heavier vehicles, it has been found that four tapered leaves can do a better job than six leaves of constant profile (and at lower cost).

Leaf springs are usually semi-elliptical in shape, flattening out under load. The top leaf is the longest and carries mounting eyes at either end.

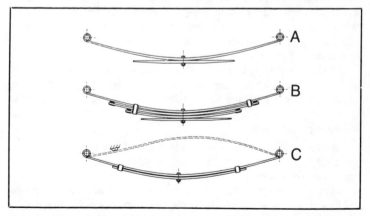

Fig. 5-2. Three types of leaf springs in general use include the single-leaf tapered-section spring (with or without auxiliary); multi-leaf spring with auxiliary leaf; and progressive-action two-leaf spring.

67

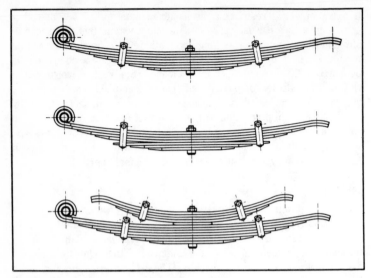

Fig. 5-3. Heavier vehicles usually have one of these types of leaf springs: (top) multi-leaf, one-stage spring; (center) multi-leaf, two-stage spring; (bottom) multi-leaf spring with multi-leaf auxiliary.

Leaves of varying lengths are positioned below, often formed with lesser curvature, the shortest leaf at the bottom coming into play only towards the end of the deflection range.

Leaf springs can be installed in many ways—transversely as well as longitudinally. Some examples from history will be given in the chapter dealing with the evolution of suspension design.

One advantage of the leaf spring is that inter-leaf friction dampens its deflections, constituting a built-in shock absorber. But that is also one of its main drawbacks, for the engineers soon found that inter-leaf friction varied a great deal according to weather conditions and lubrication. This introduces an erratic factor in their action.

Modern cars using leaf springs are equipped with wax-impregnated fabric or plastic inter-leaf liners which minimizes random variations in friction and, thereby, obtains more uniform and predictable spring action.

Coil Springs

A coil spring is made from a steel rod of circular cross section produced by winding the rod around a mandrel, usually of cylindrical shape, to form a helix.

While the leaf spring flexes under stress in direct tension, coil springs and torsion bars are stressed in shear. Coil springs can be arranged to work either in compression or in tension. On cars, they usually work in compression, standing on suspension members and carrying the sprung mass. They are wound loosely, with space for compression between the windings. The ends are usually ground flat for firm seating in abutment plates and to provide uniform deflection of the spring under load. Mounted in tension, the coil spring is usually wound in a tighter pattern and provided with attachment hooks at each end.

Cylindrical coil springs are the simplest to produce and, therefore, the most common on cars. But conical- and convex-form coil springs also exist and are used in cars where they present solutions to space restrictions or other installation problems.

Flex rates in coil springs depend on length and diameter of the steel rod. Spring rate is increased if the rod thickness is increased. Long coil springs have lower rates than short ones; the longer the rod, the lower the rate. The diameter of the winding has no effect on spring rates, but is usually chosen for convenience of installation.

Coil springs have won great favor because they can do a large amount of work in a small installation space, while leaf springs take up a lot of room. Coil springs are also lighter in themselves. They permit larger deflections than are possible with leaf springs of practical dimensions, but they are incapable of handling locating duties. They can only work as vertical load-carrying members, free of driving torque and horizontal thrust loads. Coil-spring suspension systems, therefore, must have separate control arms for wheel location.

Torsion Bars

The same is true of torsion bars. They can be thought of as straightened-out coil springs. Actually, the torsion bar can be a straight steel rod, tube or a laminated assembly of steel strips. It is loaded by twisting. One end is held firmly anchored in the chassis frame or body shell, while the other end is fixed to the pivot axis of a moving suspension member. Torsion bars can be installed in any plane, at any angle, wherever it suits the suspension geometry and fits in with the general vehicle architecture.

Coil springs and torsion bars have no inner friction (apart from the shear-strength friction in the material). Consequently, they have no built-in damping effect. Since there is no initial friction to overcome, they also work better than leaf springs in absorbing minor road-surface unevenness. In many aspects of spring action, coils and torsion bars are alike. Both can be installed to offer progressive action, though it is usually easier to do it with coil springs.

Varying Coil Spring Rates

As normally installed in a car, coil springs have a constant rate of compressibility. However, the details of installation angles and linkages can be changed to cause variations in rate, and the use of flexible bushings also has a modifying effect.

Variable rates in coil springs can be obtained by altering the wind spacing (Fig. 5-4). For instance, the upper half can be wound more loosely

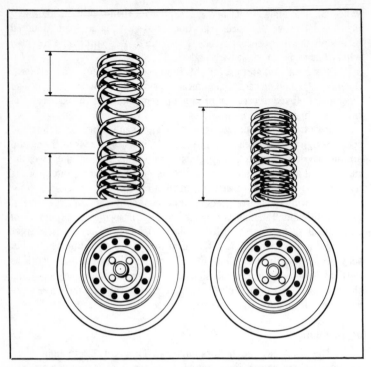

Fig. 5-4. Some measure of variable-rate action is obtained by looser winding in one section of a coil spring, which then absorbs the initial deflection.

than the lower half, or a wide-spaced section can occupy the middle area, with tighter winding towards the ends. But the amount of progressivity obtained by this means is quite limited.

A vertical coil spring attached directly to a horizontal suspension arm will work at a linear rate, directly proportional to any increase in load. By moving the anchorage point of the spring closer to the pivot axis of the suspension arm so that the spring is no longer vertical, the rates are altered. An initial load will meet little resistance and cause relatively large deflection, but as the load increases, the spring rate rises progressively and resists further deflection at a progressively increasing rate.

This does not provide a wide range of progressivity, however (Fig. 5-5). To go further, the coil spring can be placed in tension and inclined at any convenient angle, even horizontally, with an upper control arm pivot designed to move through an arc that's disproportional to the wheel deflection. With progressively greater stretch of the spring for a given amount of wheel travel, a perfectly linear-rate spring can be made to give progressive spring rates.

Similar linkage arrangements with additional cranks and pivots have been devised to work with torsion bars, notably for the front suspension of

Grand Prix racing cars where the ground clearance is very slight, since aerodynamic lift forces are best controlled if the body comes low enough to prevent air from flowing under the car, while the suspension system must prevent direct contact between the body nose section and the road surface. Progressive-rate torsion bar suspension can also be arranged for sports and family cars, though not without complication.

A British Ford patent from 1967 shows how progressive spring rates can be obtained with torsion bars. A four-link type rear axle suspension uses four transverse torsion bars. A long radius arm from each rear wheel twists one torsion bar via a rigid mounting and another via a resilient coupling. Thereby, the torsion bars are loaded at different rates.

No matter what type of metal spring is used, some provision must be made to stop deflection in jounce. At the end of rebound, nothing critical happens. But in full jounce, safety considerations dictate that a warning be given to the spring that it's approaching the limit of its travel, and that the end be marked without any shock that can damage the mechanical parts or cause serious passenger car discomfort. In other words, something is needed to prevent bottoming (Fig. 5-6). That something is called a *bump stop*, and in modern cars, the bump stop consists of a shaped rubber bumper

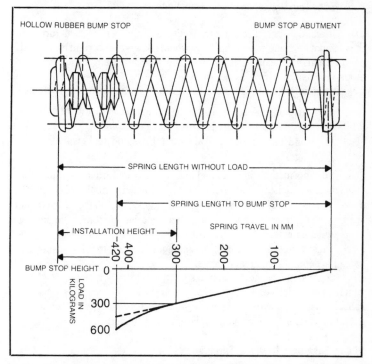

Fig. 5-5. Progressive-action coil springs can be obtained by drawing wire with a conical profile, thick in the middle and thin at the ends.

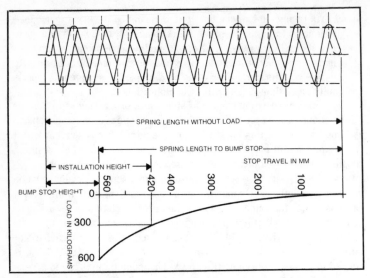

Fig. 5-6. A common means of preventing bottoming with soft coil springs is to insert rubber auxiliary springs with abutments.

that softens the impact loading transmitted to the vehicle structure. But they cannot be made so soft that they would collapse. A certain hardness is needed, and that means spring stiffness must be selected to bring the bump stops into use as rarely as possible.

Dynamic vertical forces are established primarily by three items:

- suspension rate (spring rate as measured at the wheel);
- suspension linkage and spring friction; and
- fluid damping (shock absorber tires and rubber pivot bushings).

Under normal driving on good highways, wheel travel is short (less than an inch) and the primary resistance to suspension motions is provided by friction. On the other hand, long-travel, high-impact jounce movements are resisted primarily by shock absorbers and bump stops. The sprung-mass suspension rate, therefore, is relatively unimportant in reducing the forces causing impact harshness, assuming, of course, that the suspension has an acceptable frequency range.

Suspension linkage friction has been sometimes termed "stiction", which is descriptive. In the quest to produce the greaseless chassis, rubber bushings have become one of the suspension engineer's most useful materials for improving isolation characteristics.

Shock Absorbers to Brake Springs

Shock absorbers are intimately linked with spring action and are an essential part of the total elastic element. They are dealt with fully in a chapter of their own, but it is proper to explain here the shortcomings of metallic springs that have made shock absorbers necessary.

The elasticity of a metal spring is a two-way affair. Any stress put into the spring will bring on a series of oscillations about its neutral position. Bend a leaf spring, and when released, it will bend the other way in a series of pendulum-type oscillations before going back to its static position. Compress a coil spring, and it will rebound and continue to flex in a pattern that's analogous to pendulum motion, with diminishing magnitude of travel and increasing frequency, before coming to rest.

If unchecked, such spring oscillations would continue for a time, their duration depending on the spring stiffness and the weight it supports. Think of what that would mean with the four springs on a car, each free to stretch out every bump into a series of flexings.

The two-way nature of these oscillations has bad effects on both ride and stability in a car. First, the repeated jounce motions are transmitted to the car's structure, interior and passengers. The opposite oscillations, in rebound, tend to make the wheels bounce against the road surface. Tires intermittently lose traction and sidebite if that happens. All these oscillations are cyclical and occur at definite periods.

Any weight supported on a spring has a natural or resonant frequency of bouncing. This applies to a tennis ball dropped on the ground and to the car body resting on its front and rear suspension systems. The natural frequency depends on the spring's maximum static deflection and the load on the spring. Any reduction in load or spring rate will result in a lowering of the natural frequency. Stiffening the springs or adding to the load will increase it.

It's important to know the natural frequency of a car's suspension systems, for it's a frequency to be avoided. The wheels run into bumps, or series of bumps, causing up-and-down deflections of a certain magnitude and at a certain frequency. If this frequency happens to coincide with the natural frequency of the suspension system, the springs will not absorb the bumps, but will greatly magnify the disturbance.

This is a phenomenon known as *synchronous reasonance*. You can see it by watching the swings in a children's playground—by timing his own body impulses to coincide with the reversal of pendulum-oscillation in the swing, a child can build up the swing travel to very high angles and reach magnificent heights.

Bumps which excite the suspension system at a lower frequency will also cause a rise in the magnitude of deflection. Only when the bump frequently is higher (faster) than the natural frequency, can the springs help absorb part of the wheel deflection. Because cars travel at different speeds, and roadways have no standard bump size or spacing, there is no telling what bump frequencies will be encountered. Consequently, the chassis engineer can only do one thing: Try to make sure that the natural frequency is lower than any frequency triggered by any road surface unevenness at any speed.

But not all cars are so light, nor do the engineers always have freedom to select sufficiently soft springs, to make sure that the natural frequency of a car's suspension system actually is lower than that of any disturbance encountered. The springs need something more—a sort of braking device that will slow down the cyclical oscillation and bring the whole system back to rest sooner. And that's what shock absorber is.

Ride Comfort and Resonant Frequencies

We will discuss more about shock absorbers later. At this point, we shall deal with the springs and their relationship to ride comfort (Fig. 5-7). What we vaguely describe as ride comfort is mainly an absence of disturbances transmitted to the vehicle occupants. These disturbances or vibrations, when they occur, cover a very wide frequency spectrum ranging from zero to frequencies associated only with audible noise.

With all types of suspension systems using metal springs, there are four bands of definite resonant frequencies. Up to a frequency of about 80 cycles per minute, the sprung mass will oscillate on its suspension springs and tires with minimum interference from the dampers. These are *ride motions*. At frequencies of approximately 200 cycles per minute, the sprung and unsprung mass will be somewhat locked together by friction in the suspension linkage and dampers causing "boulevard jerk."

Harshness shows up as a sudden and often persistent hard-impact noise. It is a low-amplitude vibration that occurs in the frequency band of about 200 to 1,000 cycles per minute. Between 80 and 200 cycles per minute we have *shake* (defined according to locality as beaming, torsional shake or cross-shake). That's a separate band between ride motions and harshness. *Wheel hop* occurs as a form of harshness.

At frequencies of about 600 cycles per minute, the unsprung mass may be set in oscillation between the suspension springs and the tires, resulting in the tire leaping off the road. This is *wheel hop*. During the development of a new car its wheel hop characteristics are checked out by tests on rough washboard-type roads for conditions of acceleration, cornering and braking. In modern cars, improved suspension geometry, better shock absorbers and low pressure tires have minimized the wheel hop problem.

Above 5,000 cycles per minute the vibrations are usually called *road noise*. The frequency is so high that no motion is felt, but it becomes apparent as an audible period of vibration.

Ideally, the ride qualities of a car should be soft, quiet and free from both shake and boom periods. Ride motions should be slow and balanced front-to-rear. The car should be free from lateral or diagonal cyclical motions, such as quick waddle.

A British suspension expert, David Avner, has established that low frequency disturbances cause discomfort only at relatively large amplitudes, and that the magnitude of the vibration which can be tolerated is reduced progressively with higher and higher frequencies.

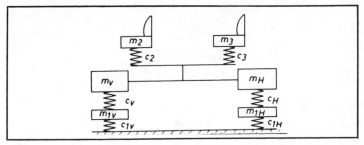

Fig. 5-7. For a true measure of ride comfort, seat springing must also be taken into account. It is the final stage of insulation between the roadway and the passengers.

When discussing ride discomfort, it becomes necessary to distinguish between *cyclical oscillation* and *random shock*. Random shock will produce no illness symptoms in the average passenger, but it is, of course, an unpleasant disturbance which suspension engineers strive to minimize.

On the other hand, human beings are extremely sensitive to cyclical motions. Low-frequency oscillations, such as ride motions of less than 100 cycles per minute, sometimes cause travel sickness. Other ride motions usually cause nothing more severe than travel fatigue (by triggering involuntary use of muscles). Shake can also have tiring effects, but vibration is mainly to be considered as an annoyance (which causes different reactions in different people).

All efforts towards greater ride comfort usually commence with an all-out attempt to control the low-frequency cyclical oscillations, and the other vibrations are later dealt with as most practical. Some vibrations can be eliminated by engineering changes; others have to be dampened by artificial means, such as deadening and isolation materials.

Numerous technical advances have been made in modern cars, with the prime objective of maximizing passenger isolation. Substantial improvements in engine mounts and engine, axle and transmission noise levels have been accomplished. Softer-riding, wide-base, low-profile tires have been developed. The use of sound insulation and deadener in passenger vehicle design has reached new peaks. Unit body construction has practically eliminated squeaks and rattles.

Accompanying a general trend towards lower spring rates, suspension developments have brought about reductions in ride friction and changes in dynamic wheel geometry which result in lower harshness levels. The use of large-volume rubber bushings in place of all-metal bushings for suspension mountings and replacing leaf springs with coil springs or torsion bars have played a large part in successfully combining metal springs with present-day ride and handling requirements.

Whether the aim is increased handling precision or increased ride comfort, the engineer has to justify his spring selection on a basis of cost, convenience, performance and reliability.

Chapter 6
Damping the Shocks

Shock absorbers are necessary to assure proper ride comfort, and they are just as necessary to help the driver control the car. What is a shock absorber? It is a velocity-sensitive, hydraulic damping device. It delivers a certain resistance to any suspension movement. Some of the energy of the spring movement is dissipated into heat inside the shock absorber and carried off by the airflow around it.

Practically all cars today use the same basic type of shock absorber; the double-acting, telescopic hydraulic damper (Fig. 6-1). Within that class definition, there are any number of variations. Damping action is based on the resistance of a fluid under pressure to flow through a restrictive valve. Any school child knows the principle from watching the water being drained from a bathtub. The drain hole has a certain size. That's the valve. The weight of the mass of water puts pressure on the water nearest the hole. But it still takes time for the tub to empty.

Telescopic Shock Absorber Construction

In the shock absorbers, the fluid is oil and the pressure is supplied by spring deflections. A whole system of valves checks the fluid flow between the chambers. Let us take a closer look at the hardware.

The shock absorber consists of two concentric tubes sealed together as a unit, a cartridge-shell type top cover, a piston and a rod, and a number of valves (Fig. 6-2). The inner tube is a pressure tube which works as a cylinder in which the piston operates. It is completely filled with fluid at all times. Its lower end is closed by a valve called the compression valve system, and its upper end is closed by the piston rod seal. The rod extends outside the assembly and carries a mounting ring at its top end.

The outer tube is a fluid reservoir whose lower end carries a similar mounting ring. It provides space for reserve fluid and for the overflow from

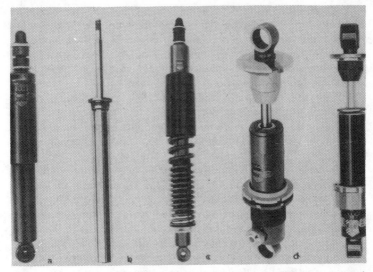

Fig. 6-1. The range of Koni shock absorbers includes special-purpose and heavy-duty dampers with two-way adjustment and auxiliary springs.

the pressure tube. The piston, rod and shell-type cover are attached to a bracket on the frame of the body. The sealed tube assembly is attached to the chassis suspension and telescopes inside the cover cylinder. This shell-type cover is a dust-shield to protect the mirror-finish of the piston rod which is usually chrome-plated and polished to a smoothness measured into micro-inches.

When the wheel hits a bump, it is deflected upwards. The piston in the shock absorber descends, pushing some of the hydraulic fluid out through the base valve and into the outer tube. Some of the fluid also passes through valves on the piston into the upper part of the outer tube. When the movement is reversed and the piston moves back up, the fluid is drawn back from the reserve cylinder and the fluid in the upper part of the main cylinder flows back through the piston valves.

Shock Absorber Operation

The double-acting shock absorber has two cycles: a compression (jounce) cycle and an extension (rebound) cycle.

The size of the valves in the piston and cylinder determines how fast the fluid flows back and forth and determines the control exerted on the up-and-down motion of the car. For uniform action, the fluid must be free of air bubbles. A helical baffle in the reservoir tube is intended to prevent air from mixing with the fluid. But in particularly jerky situations, the piston may be forced into a series of reversals of direction so quickly that the fluid flow cannot keep up and some aeration will occur.

The faster the piston moves, the greater the fluid velocity through the valves and the faster the pressure buildup. During the compression cycle, the road and piston move down in the pressure tube, which produces a small pressure drop in the upper chamber. At the same time, the volume of the lower chamber is reduced, and the pressure on the fluid contained there rises. The fluid follows the path of least resistance to correct the pressure imbalance and flows through the outer passages in the piston to begin filling the upper chamber.

Since the piston rod also displaces fluid, there is not enough room in the upper chamber for all the fluid trapped below. The remaining surplus fluid is forced under pressure through the orifice in the center of the compression valve at the bottom of the inner tube and enters the outer tube.

During the extension cycle the piston and rod move towards the top end of the pressure tube and the volume of the upper chamber is reduced. This now becomes a high-pressure tank. To equalize the pressure, fluid flows down through the piston valve system into the lower chamber, but the fluid arriving by that route is insufficient in volume to fill the lower chamber (because of piston-rod displacement). That causes a pressure drop in the lower chamber which forces the compression valve to unseat, thereby admitting fluid flow from the reservoir in the outer tube.

All modern shock absorbers are made with three stages of valving. The valve functions are separated according to suspension movement and need for damping. The first stage is a low-velocity valve, designed to resist initial pressure buildup, as during body roll on a curve. The size of its orifice is a compromise, for a large one adds nothing to roll resistance, while a small one can cause a harsh ride.

The second stage is a blow-off valve, triggered by medium piston velocity, controlling the connection to the reservoir. After this stage is activated, the first stage has no significant affect any more.

The third stage is a final restrictive passage, activated only at maximum pressure and peak piston velocity. Due to the second stage blow-off valve, the third stage has no affect on shock damping at lower pressure rates. But with a wide-open blow-off valve, the final restrictive passage becomes the dominant factor in damping.

One of the problems with such designs is that the fluid heats up as it is forced through the valve constrictions. As its temperature rises, the viscosity drops, and it begins to flow more easily than when cool. Thus, the damping rate is not consistent. Of course, chemists have developed fluids with a high viscosity index, less sensitive in their flow characteristics to changes in temperature.

Shock Ratios and Overall Damping Rates

All valves are two-way valves, so that they function in rebound as well as in jounce, but the damping characteristics are not necessarily the same both ways (Fig. 6-3 and 6-4). Each shock absorber has a shock ratio such as 50/50, 75/25, 10/90, and so on. With a 50/50 shock ratio, the damping is equal in jounce and rebound. Which is which? Jounce is always mentioned

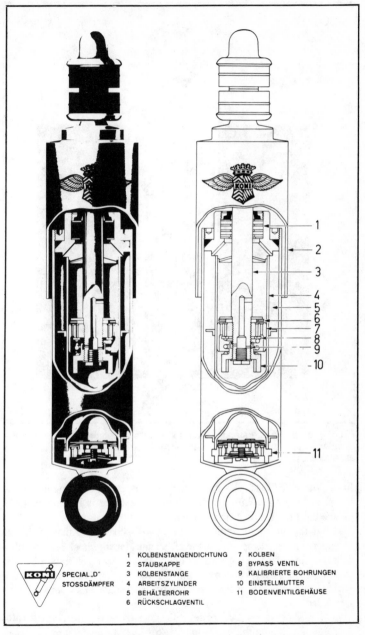

	1	KOLBENSTANGENDICHTUNG	7	KOLBEN
	2	STAUBKAPPE	8	BYPASS VENTIL
SPECIAL „D"	3	KOLBENSTANGE	9	KALIBRIERTE BOHRUNGEN
STOSSDÄMPFER	4	ARBEITSZYLINDER	10	EINSTELLMUTTER
	5	BEHÄLTERROHR	11	BODENVENTILGEHÄUSE
	6	RÜCKSCHLAGVENTIL		

Fig. 6-2. Inside the Koni shock absorber: 1 piston seal; 2 dust cover; 3 piston rod; 4 working cylinder; 5 outer tube; 6 return valve; 7 piston; 8 bypass valve; 9 calibrated apertures; 10 adjusting nut; and 11 lower valve body.

79

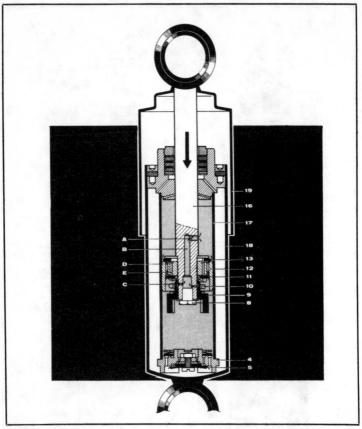

Fig. 6-3. In the jounce stroke, the piston moves to compress the fluid in the cylinder, which is then forced into the reservoir cylinder via the foot-valve assembly.

first. With a 75/25 shock ratio, three-quarters of the damping lies in the compression cycle. With a 10/90 shock ratio, there is little damping in jounce but practically all in rebound.

But the shock absorber with the 75/25 shock ratio may still have stronger rebound damping than the 10/90 unit, for the shock ratio is only a relative expression of the distribution of damping control. Overall damping rate is measured in pounds of control force over the full piston stroke at various velocities which correspond to typical frequencies where damping is needed.

Most shock absorbers are set with shock ratios that provide a greater share of damping on the extension stroke—not because it's more essential, but because less of it can be tolerated on the compression stroke. Jounce deflections are already resisted by the spring rate (and any additive stabilizer bar effect), so to bring a high damping force into action in that situation would certainly result in a harsher ride.

Rebound damping, on the other hand, gives greater benefits. It tends to stabilize the car in its normal ride motions by restricting the return oscillation in low-frequency deflections. It delays the wheel's dropping into ruts or potholes (resisting the spring force) and, thereby, smoothens the ride. And if the wheel has actually been kicked off the road surface by a bump, rebound damping will let it down slowly and reestablish traction and sidebite while preventing renewed bouncing.

Shock Absorber Location

The amount of control force needed in a shock absorber is dependent on many factors. We have talked about the sprung-to-unsprung weight ratio, spring rates and frequencies of oscillation. The suspension linkage

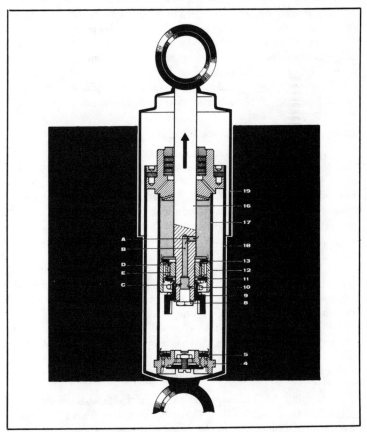

Fig. 6-4. On rebound, the piston relieves pressure in the inner tube, and the piston valve admits fluid to pass through it to the low-pressure area.

or, to be exact, the location of the shock absorber's attachment point on the suspension member, relative to the length of the suspension member and the distance from the wheel hub to the shock absorber mounting, also plays a part. If the shock absorber is mounted close to the wheel hub, the piston stroke will correspond closely to the wheel travel for any given bump. But it is often not practicable to install the shock absorber next to the wheel hub.

Usually it is mounted near the mid-point of the suspension control arm (halfway between the ball-joint and the pivot axis in a modern front suspension system). In this example, piston stroke will be about half as long as the wheel travel. That corresponds to a 2:1 link ratio. The relationship between wheel travel and the required control force in the shock absorber rises as the square of the link ratio. A car with a 1.8:1 link ratio needs damping characteristics with 36 percent higher strength than a car of similar weight and spring rates but a 1.2:1 link ratio.

The shock absorber is designed to cover a wide range of oscillation frequencies, from one or two to upwards of 500 cycles per second. It is in the nature of hydraulics that high-frequency oscillations are most easily damped because of the time element involved in any fluid flow. But one of the most important oscillations the shock absorbers have to damp occurs at relatively low frequency and tends to increase in amplitude. This is the natural frequency of the unsprung mass, which usually occurs between 10 and 18 cycles per second.

Adjustable and Gas-filled Shock Absorbers

Most shock absorbers have no adjustment possibilities. Either their damping characteristics are suitable for the car, or they are not. Adjustable shock absorbers have been on the market for many years, however, and have important advantages. (Fig. 6-5) Their rebound damping can be altered by a simple operation (without tools and without removal from the car). Racing-type shock absorbers also have compression-cycle adjustments, but for cars in road use, rebound adjustment is generally all that's needed.

Adjustable shock absorbers also permit their original damping characteristics throughout their life. Standard shock absorbers must be junked when worn. What wears them out? Friction-millions of jerks and jolts rub the parts and seals to the point where leaks are created.

In adjustable shock absorbers, some leakage can be counteracted by resetting the unit for higher control force. Adjustable shock absorbers have parts and seals that wear out too, but they can be rebuilt at relatively low cost. Non-adjustable shock absorbers are sealed and must be replaced as a unit.

Many high-performance cars are equipped with what are called *gas-filled shock absorbers* (Fig. 6-6). They differ dramatically from the types we have dealt with up to now. Instead of one piston and two tubes, they have a single tube and two pistons. The lower piston is the working piston, carried at the upper end of a piston rod and equipped with hydraulic valves.

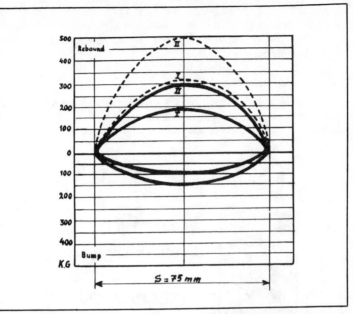

Fig. 6-5. Dotted lines represent, the full range of rebound resistance adjustment in a Koni shock absorber, while the arched solid lines show the standard setting. The solid lines below the zero-line represent jounce travel resistance.

The upper piston is a separator piston which divides the chamber above the working piston into two parts. The space below the separator piston is filled with oil. Above the separator piston is a chamber filled by an inert gas, usually nitrogen, under high pressure, 25 bar or 368 pounds per square inch. Under extreme operating conditions both fluid and gas temperature will rise, and gas pressure may double (to 50 bar or 736 psi).

The gas pressure serves to keep the oil-filled chamber above the working piston at minimum volume at all times, without interfering with the damping characteristics. As long as the separator piston provides an effective seal, aeration and foaming in the fluid are eliminated. That is perhaps the most important advantage of gas-filled shock absorbers.

The gas is compressible, while the oil is not. The gas has a second task, which is to yield some extra volume back to the oil chamber as the oil expands due to the heat generated when the shock absorber is working hard. The excess fluid has no other place to go.

The chamber below the working piston is also filled with oil. It is sealed at the bottom, so that all fluid displacement takes place between the two chambers on either side of the working piston. The piston has a seal ring on its outside surface, but it is made with something less than absolute sealing effectiveness. About 10 percent of the fluid displacement takes the route along the tube wall outside the piston, while 90 percent is forced

83

through the piston valve system. This serves to lubricate the piston and
tube wall surfaces to minimize wear and also to provide a safety valve,
preventing the blocking of spring deflection on sudden high-amplitude
suspension movements.

Coil Spring Load Levelers

All shock absorbers we have dealt with up to now are strictly dampers. They can be combined with auxiliary springs, however, so as to

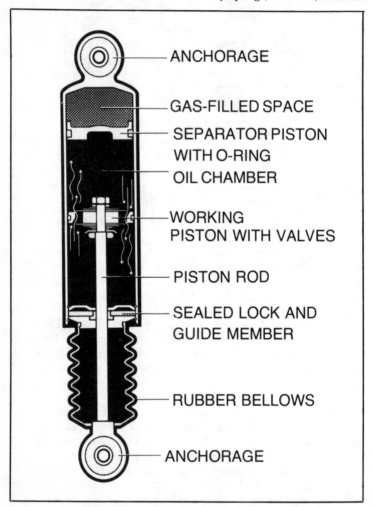

Fig. 6-6. This section of a Bilstein gas-filled shock absorber shows gas compartment above, with free separating piston between it and the hydraulic portion. Foot-valve assembly has been replaced by a seal-and-guide assembly.

participate in the load-carrying duties of the suspension system and form a unit that is usually called a *load leveler*. The auxiliary spring can be a coil surrounding the shock absorber tubes, or a gas spring positioned at one end of the shock absorber.

The coil-spring type of load leveler adds to the spring rate and, therefore, also the roll stiffness. It can be an enormous help for cars with low ride rates and excessive suspension amplitudes. It can also detract from the proper and necessary spring deflections on some cars and must, therefore, be installed with a good deal of discrimination.

The existence of the auxiliary coil spring reduces the demand for jounce damping in the shock absorber, but adds to the need for rebound damping. Load levelers are mainly used in the rear suspension system and less commonly for the front wheels, though special front-wheel load levelers are on the market.

Delco and Koni Level Controls

The most common type of auxiliary gas spring combined with hydraulic dampers is what Delco calls its *Superlift* shock absorber (Fig. 6-7). It is a normal telescopic unit with a pliable nylon-reinforced neoprene boot acting as an air chamber. The unit will extend when inflated and deflation will make it retract. The two shock absorbers for a rear axle are interconnected via a flexible air line that equalizes pressure in the two air chambers. Air pressure of eight to 15 psi is maintained in the air chambers at all times in order to minimize boot friction. This is accomplished by a check valve in the exhaust fitting on the control valve. The system is so designed

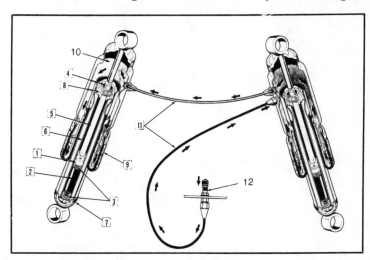

Fig. 6-7. Delco Superlift shock absorbers have air chambers on top (10) connected to fill valves (12), which in turn are hooked up to compressed-air reservoir.

that in the event of air pressure loss, shock absorber function and the action of the conventional springs will continue without being impaired. Compressed air is supplied from a reservoir that can be filled at any service station (from the tire-inflation air hose).

The most sophisticated combination of a hydraulic damper and suspension spring is a unit made by Koni, consisting of two gas springs connected in series plus damping devices that are sensitive to both spring mass and fluid displacement velocity. This Koni hydro-pneumatic suspension unit includes valves for automatic level control and is, in fact, intended as the sole means of suspension. It constitutes a cross-over point between conventional spring-and-damper combinations and a full four-point, self-leveling suspension system, and will be dealt with in a later chapter.

Chapter 7

Independent Front Suspension Systems

We have discussed front axle suspension systems at sufficient length in earlier chapters. As pointed out, they are no longer used for cars, except certain four-wheel drive, off-road vehicles, where they are preferred for reasons of ground clearance, suspension travel and ruggedness. Normal road cars have gone to independent front suspension. That means that each front wheel is free to make jounce and rebound movements without affecting the other one. It also means doing away with the axle and rigging up a system of pivoting control arms that locate the wheel, liberating the springs from any form of locating duty. There are many ways of doing this. Control arms can be arranged in many planes from trailing to transverse, vary greatly in length and arc and work with a number of different types of springs.

What all types of independent front suspension promise can be summed up in a short list: reduced unsprung weight, restricted body roll without detriment to ride comfort, minimized gyroscopic effects in the steering, and separation of the flexible element from the locating duties, opening the way for the use of softer springs in combination with more accurate wheel guidance. Depending on the type of system and the detail design and specifications, they fulfull their promises to a smaller or greater extent.

I have chosen to group the systems not according to type of springing, but on the basis of the suspension linkage, which is what dictates the geometry.

Twin I-Beam

The simplest of all independent front suspension systems is known as Twin I-beam (Fig. 7-1). Though it is not actually used on any current

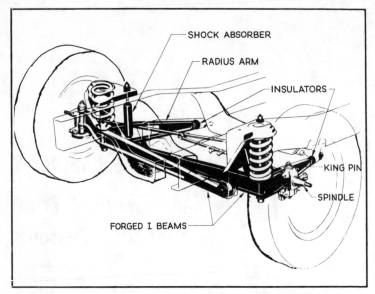

Fig. 7-1. Twin I-beam suspensions for Ford trucks used leading radius arms and Haltenberger steering linkage.

production car, it is a feature of Ford's light-duty trucks. Here, each front wheel has what begins as its own axle. The opposite end, however, instead of carrying the other wheel, is anchored in the vehicle frame.

Each of the two beams is longer than half the track, so that they overlap and must cross each other in the middle of the chassis. This is easily done by staggering the anchorage points—moving one forward and the other backward relative to the front wheel axis.

The same type of leaf springs can be retained, with the same drawbacks and advantages, because they would still perform a locating duty. Ford's trucks have coil springs, standing straight up on the beams, which are located by leading *radius arms*. That's a term which needs to be defined. A radius arm is an arm that is hinged so that when pivoting, the arm's other end describes an arc, always with the same radius. It is a rigid arm that does not stretch or contract, flex or bend.

When such an arm is said to be leading, it is positioned behind the wheel whose suspension system it forms part of, but ahead of—that is, leading—its pivot point. The opposite of leading is trailing. A trailing arm trails from its pivot point to the wheel. Think of it this way. With trailing arms, the wheels act as trailers, positioned behind the suspension linkage.

The Twin I-Beam system shares some of the advantages of the axle. It has a high roll center, though not as high as the spring anchorage points. To find its roll center, you draw a straight line from the I-beam anchorage point in the chassis frame to the intersection between the center line of its wheel and ground level. Draw those lines for both wheels. Where the lines cross is your roll center. In the case of the Ford truck, it's 6.8 inches above ground level, while a truck with front axle suspension using the same size wheels and tires would have its roll center 12.8 inches above ground level.

Twin I-beam suspension has considerably less unsprung weight than the axle, the beams themselves counting partly as sprung weight (nearest the chassis mountings) and partly as unsprung weight (at the wheel end). The same applies to the radius arms and the springs.

While a single-wheel bump will not affect the other wheel, the system permits significant camber changes in roll. Because the steering axis geometry is fixed by the angle of the king-pin bearings in each beam, and the beams make a pendulum-type swinging movement from their pivot points, camber angles vary according to wheel deflection—negative in jounce (upwards deflection). This introduces variations in track, which are undesirable for dynamic stability and cause tire wear.

Since body roll is occasioned by lateral weight transfer, it causes spring deflection. When the outside spring is compressed, the inside one is stretched. For five degrees of body roll, the wheels on a Ford truck change more than two degrees of camber (in opposite directions) on the front wheels.

The negative camber on the outside wheel is helpful in improving sidebite and also in aiming the tire more sharply into the turn, while the positive camber on the inside wheel assumes a more or less parallel angle and gives an added benefit (though smaller, because it carries less of the load).

The camber change in the wheel that goes over a bump corresponds closely to the effect of a bump on one pair of axle-mounted wheels.

Horizontal compliance is ample if longitudinal lead springs are used. The Ford System, with radius arms, has no more horizontal compliance than is provided by the flexibility of the rubber bushings of the radius-arm pivot bearings.

Trailing Links

Trailing links (Fig. 7-2) are like the trailing arms in the sense that they hold the wheel in a trailing position at the rear end of the links. What is now called *trailing link suspension* is a very specific type of design in which each front wheel has two trailing links of equal length, one above the other, carrying the wheel hub via ball joints or the upper and lower end bearings on the king-pin.

The front ends of the links, which are short and slim members made of high-strength steel, are pivoted on shafts mounted in the chassis. In most cases, the trailing arms do not conform exactly to the car's center line, but run at an angle so that the ball joints are spread farther apart than the pivot points on the chassis. They remain pure trailing arms in their arc motion as long as the pivot shaft runs in the exact transverse plane, regardless of this apparent deviation in their shape.

Fig. 7-2. Trailing-link suspension for the Volkswagen Beetle used transverse torsion bars connected to both upper and lower sets of trailing links.

Usually the trailing arms are angled downwards at the rear in the car's static position so that initial jounce deflection gets the benefit of some degree of horizontal compliance (which, however, is reversed towards full jounce travel). The longer the trailing links, the greater the amount of horizontal compliance. They are usually short enough to keep changes in wheelbase within a fraction of an inch.

The wheels maintain their static camber settings on single-wheel bumps, and variations in track are completely avoided. But this is not to say that trailing links assure constant camber, for body roll produces an identical tilt in the wheels. This is forced on the wheels by the unavoidable displacement from the horizontal plane of the pivot shafts which are mounted on the chassis frame. Thus, on a right-hand curve, the outside wheel is forced into a positive camber angle while the inside wheel assumes a parallel negative camber angle. Consequently it becomes extremely important to restrict body roll in cars with trailing links in the front suspension.

Unfortunately the system is inherently weak in terms of roll stiffness. The roll center is located at ground level (Fig. 7-3). The demand for a large-diameter stabilizer bar is difficult to satisfy because the flex rate of the bar is additive to the spring rate, and therefore threatens to upset ride comfort.

Unsprung weight is very low in trailing-link front suspension systems. Only the wheel assemblies are totally unsprung, for the springs are carried as sprung weight, and the trailing links themselves are at least 50 percent sprung weight. But trailing-link systems do not lend themselves to the provision of anti-dive effect, being designed so that the links do their normal work in the fore-and-aft plane.

Because their directional aim coincides closely with the direction of braking thrust, the trailing links are not subjected to heavy bending or twisting loads under braking. Side forces exert considerable stress, however, but with adequate width in the pivot bearing and fairly short links, they can still be made very light and slender.

The worst problem with trailing link suspension lies in the pattern of wheel deflection. Each wheel has an instantaneous center of gyration, and bumps can set up severe gyroscopic reactions in the wheels that are transferred to the steering wheel—a phenomenon known as *wheel fight*. The bigger the bumps, the greater the speed and the higher the steering angle, the worse the problem becomes.

Trailing links can be combined with almost any types of spring. The best known cars equipped with trailing-link front suspension are the Volkswagen Beetle (Fig. 7-4) and Porsche Type 356. Both used torsion bars, mounted transversely and effectively forming pivot shafts for both the upper and lower pairs of trailing links. In this system, the torsion bars also acted as stabilizer bars.

The normal stabilizer bar as fitted on most popular cars today is, in fact, a torsion bar running across the chassis and linking the two wheels on opposite sides. It is bent near the ends to form trailing levers, while the center section is firmly anchored (but free to rotate) in the chassis. The trailing levers arc up and down with wheel deflection so that on a single-wheel bump, the bar is twisted and the jounce travel resisted. On a

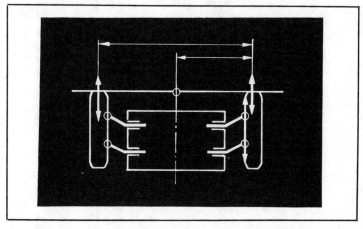

Fig. 7-3. Trailing links give a roll center at ground level, since the instantaneous center is at infinity. Body roll can move the roll center sideways, but not vertically.

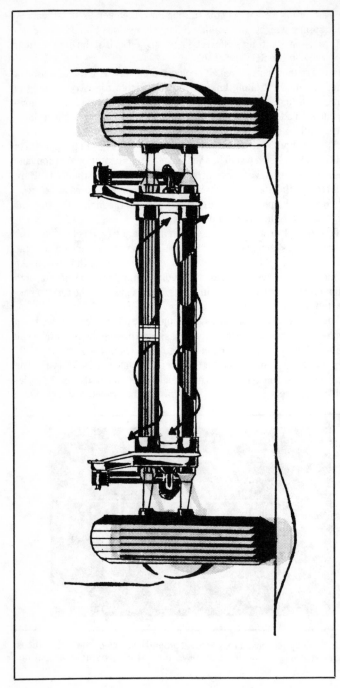

Fig. 7-4. Volkswagen used laminated torsion bars enclosed in steel tubes. Though trailing links ran diagonally, their axes were transverse for pure trailing motion.

two-wheel bump, however, the two ends make identical movements and the bar simply turns accordingly, but does not twist.

Body roll forces the opposite ends of the stabilizer bar into arcs in opposite directions since the outside wheel goes into jounce and the inside wheel into rebound. This maximizes the twisting of the stabilizer bar and the resistance to roll. In the VW Porsche setup, the torsion bars do double duty as springs and stabilizer bars (Fig. 7-5).

Trailing links have also been combined with coil springs. Notable examples are the Aston Martin DB 2 and Healey Silverstone. The upper pivot shaft was provided with a bell-crank which compressed a coil spring mounted vertically inside a box or cylinder. On the 1950 BRM V-16 trailing links were combined with oleo-pneumatic struts, using a similar crank arrangement. In these cases, separate stabilizer bars were needed to restrict roll. Trailing links also could be combined easily with a transverse leaf spring, but no such design has been used on a production model.

Leading-arm front suspension is not exactly the opposite of trailing link design, for existing examples of leading-arm types use a single arm to carry the wheel hub. Current production cars that use the leading-arm type of front suspension system are all in the same family of Citroen models derived from the 2 CV. Since they also use interconnected front and rear springs, the system will be described in a later chapter.

Transverse A-Arms

The most popular types of independent front suspension come within a vast category that can be characterized as *transverse A-arms*. To get the picture, imagine two metal frames shaped link the letter A, placed more or less horizontally, one above the other, with the bases of the triangles mounted on pivot shafts on the chassis and the tips holding the ball joints of the steering knuckles (Fig. 7-6).

The A-arms are control arms. Because of their shape, the British call them "wishbones." They are not always A-shaped, but usually form a

Fig. 7-5. Near the end of the period during which trailing arms were used, Volkswagen added a stabilizer bar and went to ball-joints instead of link pins.

93

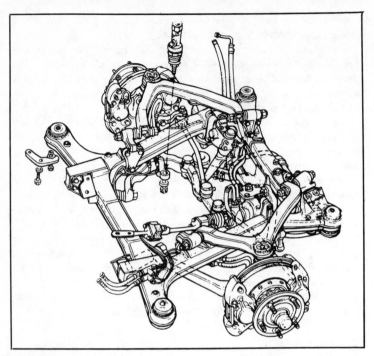

Fig. 7-6. Rolls-Royce Silver Shadow has wide-based upper and lower A-arms. Coil springs are not visible here.

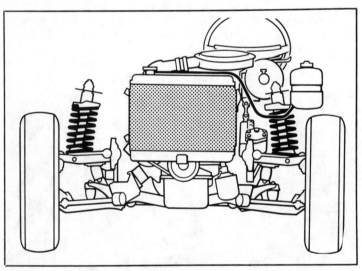

Fig. 7-7. Typical front suspension layout with unequal-length A-arms and coil springs standing on the upper control arms as used by Fiat on the 132 series.

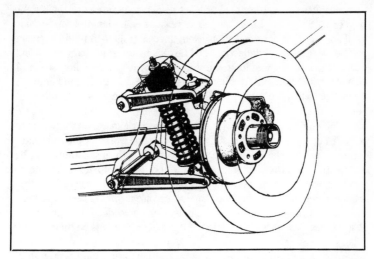

Fig. 7-8. Toyota 2000 of 1966 had unequal-length A-arms with narrow-diameter vertical coil springs enclosing telescopic shock absorbers.

triangle between the wheel hub and the chassis frame. They can be one-piece steel forgings or stampings, or they can be assembled from a number of parts.

What they provide, when the car is viewed head-on, are four firm location points for the sides of a trapeze. Such linkages provide exact and positive guidance of the wheels at all times and the length of the control arms and the height of their pivot shafts are the principal factors that determine the suspension geometry.

The two A-arms per wheel on modern cars are not equal in length (Fig. 7-7 and 7-8). The upper one is invariably shorter than the lower one.

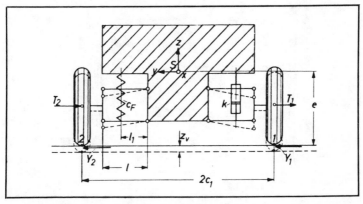

Fig. 7-9. With equal-length A-arms, wheels remain vertical on single-wheel bumps, but there are minor track variations.

Some manufacturers refer to their front suspension systems as the 3SLA1 type (SLA Short and Long Arm). The reasons equal-length arms are not used are easily understood. Equal-length arms (Fig. 7-9) tend to remain parallel during wheel deflection, and this keeps the static camber setting on the wheel throughout its travel. But it gives the same problem with roll camber that exists with trailing arms—the wheels change camber angle according to body roll, giving positive camber on the outside front wheel (Fig. 7-10).

The engineer is faced with the same need to add external means to provide adequate roll stiffness. This is often made worse by the roll center location, as we shall soon see. With equal-length arms, there can be substantial variations in track. Each wheel goes into its widest-track position when the A-arms are horizontal, and then moves inboard in either jounce or rebound. On the other hand, there are no wheelbase variations, except for what little horizontal compliance is provided by flexing of the A-arms themselves and deflection of the rubber bushings on their mountings.

With a long lower A-arm, track variations are minimized; a short upper A-arm opens the way for controlling camber variations during wheel deflection. Roll camber effects can be counteracted by careful design. Take an example where the upper control arm is two-thirds the length of the lower one. Say they are parallel when the car is at design height, and the wheel is set with zero camber. The upper ball joint describes a tighter arc during deflection than the lower one and, as a result, the wheel tends to go into negative camber in both jounce and rebound.

Leave the ball joints and lower arm where they are and bring down the upper A-arm pivot shaft an inch. The result will be to send the wheel into positive camber towards full rebound, while exaggerating the negative

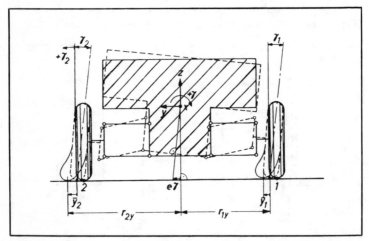

Fig. 7-10. Body roll plays havoc with the geometry of equal-length A-arms, as camber angles follow the roll angle.

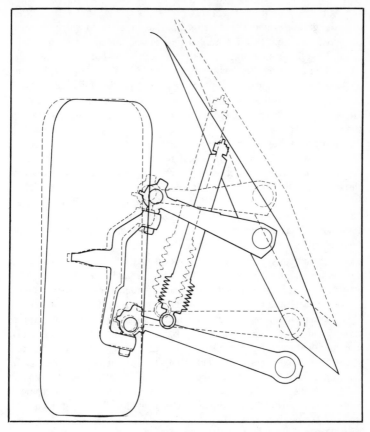

Fig. 7-11. Alfa Romeo provides an example of how body roll can put the outside front wheel into a negative-camber angle (solid lines) thanks to use of unequal-length A-arms. Dotted lines represent static position.

camber in jounce. Raising the upper A-arm ball joint will have the opposite effect. The wheel will tend towards positive camber in jounce and an exaggerated negative camber in rebound. Shortening the upper control arm increases the camber change relative to wheel travel and presents a barrier to the extent of travel that's possible with the design. Shortening the lower arm tends to diminish camber changes relative to wheel travel, but causes an increased tendency towards variations in track.

Factors in Control-Arm Design

So far, it's simple enough. But the engineers have to keep track of changes in caster and toe-in as well when selecting the dimensions and pivot points of the control arms. Most steering linkages are so designed that wheels tend to toe-out in jounce and toe-in on rebound. Since it's always the outside front wheel that carries most weight and, therefore, does the greater part of the steering job, that's the one to which we must pay most attention. Positive camber plus toe-out means stronger under-steer. Negative camber can cancel the effect of toe-out (Fig. 7-11).

The next thing they have to consider is *roll center location*. It's the angle of the control arms (with the car viewed head-on) that determines the height of the roll center. Calculating roll center height is really a very simple matter. There are just a few straight lines to trace, once you know the points that count.

Take an example: The lower A-arm is horizontal (Fig. 7-12), and the upper one slopes from a high pivot shaft to a low ball joint. The length of the control arms does not matter when we are looking for the roll center—only the angles. Extend the lines from both control arms outside the car. They intersect at a point which is called an *instantaneous center*. It's only an abstract point, but we need it for the next line to be drawn. Draw a straight line from the instantaneous center to the center of the tire footprint, and beyond, till it crosses the center line of the car.

This line, as you can see, goes underground inboard of the tread. That means you end up with a roll center below ground level (Fig. 7-13). Its exact location is the point where the line extended from the instantaneous center through the footprint intersects the car's vertical center line.

Try a design where the upper control arm slants the other way (pivot axis lower than the upper ball joint). Extend the lines from both A-arms, not outside the car, but into the car. They cross halfway between the car's center line and the opposite wheel. That's the instantaneous center. Draw a straight line back from there to the middle of the tire footprint. This line intersects the center line just a bit lower than the extended line from the lower control arm, and the roll center is located well above ground level (Fig. 7-14).

Thus we see that the suspension setup that gives the greatest nega-tive camber angles in jounce also offers the highest roll stiffness. This is the kind of geometry you find on sports and racing cars.

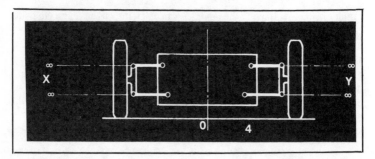

Fig. 7-12. Roll center at ground level results from keeping unequal-length A-arms horizontal. Still, this design gives better control over track and camber than equal-length arms.

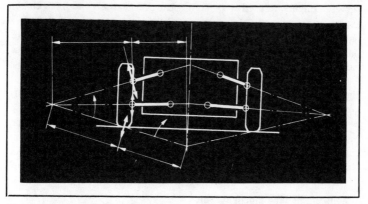

Fig. 7-13. A sloping-to-outboard attitude on the upper control arms drops the roll center below ground level if lower control arms are allowed to remain horizontal.

Finally, take the case of the parallel and horizontal control arms. Their extended lines never meet, so the engineers say the instantaneous center is at infinity. They draw a line back from infinity through the tire footprint, and find the roll center located exactly at ground level. That's clear. No problem. But the roll centers we have calculated are only theoretical. A car has a dynamic life, and the angles of the control arms keep changing as it moves along the road surface and through curves. Consequently, the roll center does not always remain in its proper (static) place, but varies its position according to spring deflections.

Body roll can cause the roll center to move up or down or far to one side as the center line moves away from vertical, according to the details of the exact geometry selected.

Its exact location during typical cornering situations is of the greatest importance, for that's when handling precision is most needed. Cars at standstill are safe from going out of control no matter where their roll

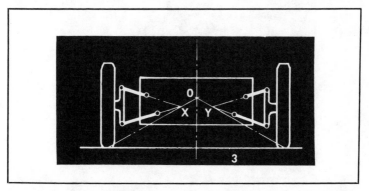

Fig. 7-14. A high roll center results by moving the pivot axes close together But this angle on the lower control arms leads to positive camber in rebound.

centers. Engineers used to make intricate calculations and trace charts more or less accurate show the interaction between all these factors. Nowadays, they just feed all the data into a computer which, in a fraction of the time it would take an engineer, draws neat pictures of everything that can happen in the suspension system.

Few cars using transverse A-arms are made with roll centers so high above ground level that the front suspension provides adequate roll stiffness in itself (Fig. 7-15). But there are some. The Saab 99 and 900 are designed with roll centers about 5.0 and 5.5 inches above ground level and do not need stabilizer bars. The softer the springs and the lower the roll center, the greater the need for stabilizer bars; but their installation is not without side effects.

Stabilizers Reduce Independence

When stabilizer bars are fitted between the two front wheels of a car, the suspension system is no longer fully independent. This form of interdependence has the beneficial effect of reducing body roll in return for the disadvantage of adding to the ride rates of the springs themselves on single-wheel deflections (Fig. 7-16).

Take the example of the Jaguar XJ-6, in which the springs give a front wheel rate of 85 pounds per inch in a single-wheel bump. The effect of the heavy stabilizer bar increases this rate to 182 pounds per inch.

Transverse A-arms, having longitudinal pivot axes, tend to provide good control of gyroscopic reactions in the wheels. As for unsprung weight, transverse A-arm designs are closely comparable with trailing-link systems. The latter has the edge, but it can be a very small edge.

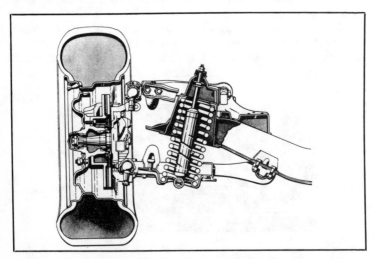

Fig. 7-15. Opel Diplomat of 1969 had inboard tilt on coil springs and shock absorbers. Roll center was close to ground level.

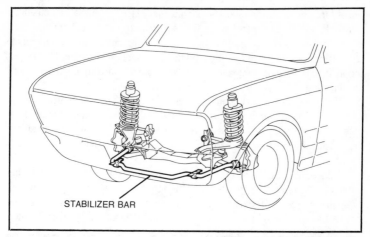

Fig. 7-16. Stabilizer-bar installation on Fiat 132 shows how it's unaffected by a two-wheel bump, acting only in response to body roll or single-wheel bump.

A narrow-splayed "A" profile gives the greatest springiness in the arm itself, while a wide-based "A" distributes the brake thrust (and other directional forces) over a wider area and, is therefore, less likely to give fatigue failure.

On some cars, the lower control arm is separated into a transverse I-arm and a diagonal drag strut. The I-arm pivots at a single point, not along an axis, so that it offers no resistance to fore-and-aft forces. The drag strut is a straight rod anchored in rubber bushings at each end, and the flexibility of these bushings provides the required horizontal compliance.

Articulated drag struts have been used but have been given up again. Some cars use compression struts (Fig. 7-17), which are drag struts turned around from a semi-trailing to a semi-leading position, working in compression instead of in tension. They do essentially the same job, again, due to the rubber bushings.

Rubber Does Not Compress

It has become common practice to insert rubber bushings in all suspension member mountings and joints, front and rear. They aid materially in restricting noise and harshness, provide a measure of flexibility and are maintenance-free (eliminating metal bearings and grease points). Of course, rubber does wear out, and bushings may have to be replaced in the car's lifetime.

Contrary to what most people believe, rubber is incompressible. Its visible flexing consists in a change of shape, but not in a change of volume. If pressure is put on a rubber bushing so as to shorten it, the result will be a growth in its diameter. In such a situation, the bushing expands radially outward, while contracting in its axial plane. Its radial swelling assures equal force distribution over the bushing's surface. It works as a cushion whose flexibility is controlled by the material strength, bushing size and shape, pre-stress loading and flange design. A *flange* is a lip at each end of the bushing. In some applications, no flange is needed. The shoulder stops axial "flow" in the bushing and plays a key role in controlling axial load on the bushing.

Bushing design has become a science in itself. They come in a variety of shapes and combinations of materials with different flexing characteristics. Most are solid, but other types are voided. Voiding of a bushing means forming the bushing with hollow areas so as to provide an extra local flexibility.

Since rubber bushings came into use for control-arm pivot points, the engineers have had to pay more attention to the deflection-steer factors in suspension design. Suspension systems with rubber bushings leave the control arms a certain resiliency. They are deflected in the fore-and-aft

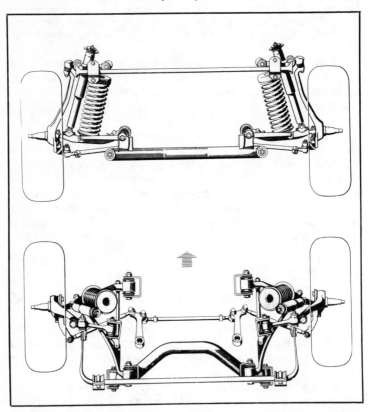

Fig. 7-17. Mercedes-Benz front suspension seem from the front (top) and from the above (bottom). Lower arms are a composite of drag strut and compression strut, with an extremely wide base.

plane by the horizontal component in bump forces, as we have seen. They also deflect slightly when exposed to side forces and the aligning forces acting on the tires in a curve. These changes in suspension attitudes describe the deflection-steer characteristics of a car.

Two Theories on Scrub Radius

Recent trends in Europe and America seem to go in opposite directions, though with similar goals. General Motors believes in a positive scrub radius for cars with rear-wheel drive as a means of assuring high directional stability without creating problems in turning effort at low speeds, and to eliminate excess camber without running into tire wear problems. GM's front-wheel-drive cars have a negative scrub radius for stability in braking, especially on surfaces with uneven friction coefficients left and right.

Buick's Accu-Drive, introduced on the 1969 models, reversed the camber change geometry of previous models (and has since been adopted by most GM cars). It is a compromise solution that gives greater stability in single-bump situations when traveling down the straight, but at the expense of increased roll understeer.

Accu-Drive cars, jounce travel puts the wheel into a slight positive camber angle instead of assuming a rather high negative camber, as on earlier models. This was obtained by moving the control-arm pivot axes farther apart, the upper one upwards and the lower one downwards, completely altering the static positions of the control arms.

With Accu-Drive, the center of the tire footprints moves fractionally closer to the center of the car when hitting a bump. On older cars, the footprint center moved outwards with increased bump deflection. This generated a force which showed up as a camber thrust acting towards the center of the car and tending to steer the car away from the bump. Newer models are practically free of this steering phenomenon.

In Europe, there is a noticeable trend towards eliminating the scrub radius. Mercedes-Benz has gone to center-point steering on all its cars, with a front-wheel suspension system that is worth a closer look (Fig. 7-18 and 7-19). The lower control arm is triangular in shape, made from three forgings riveted together, has front and rear pivot points well spaced out and is equipped with large rubber bushings.

The lower ball joint is located practically at hub level, with a tall extension that places the upper ball joint above the highest point on the tire. The upper control arm consists of a transverse I-link and a leading arm that is actually an extension of the stabilizer bar which crosses the chassis behind the front wheel axis and steering linkage.

The swivel axis has been inclined so as to coincide with the center of the tire footprint. At the same time, a very high caster angle was adopted. This gives better straight-line stability, and the steering works with greater accuracy. Cornering stability is improved and response time shortened (which means less understeer, in contrast with the GM setup).

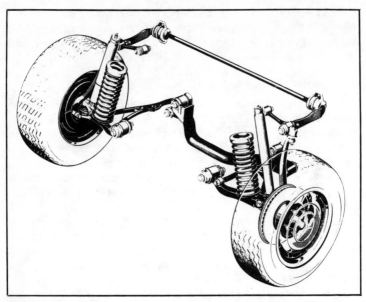

Fig. 7-18. Mercedes-Benz front suspension has stabilizer bar emerging as part of upper control arm, with coil springs standing on the lower control arms, separate from the shock absorbers.

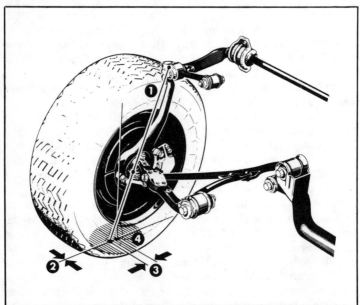

Fig. 7-19. Front suspension on Mercedes-Benz sedans was developed from the C-111 and features center-point steering (zero scrub radius).

104

The adoption of center-point steering also assures absence of steering effects when traversing single-wheel bumps at speed on straight roads. Steering stability is not affected by tire damage, uneven braking or other external causes. The Mercedes-Benz design allows ample horizontal compliance without loss of precision in the wheel guidance and provides progressive anti-dive control.

When braking, forward weight transfer occurs due to the deceleration, and the car will tend to nose-dive unless the front suspension has an instantaneous center (in the longitudinal plane) that coincides with the neutral line for the deceleration force. This can be done by proper selection of control-arm geometry. But if the instantaneous center is higher or lower, the deceleration force will have an arm on which to form a moment, applying a torque about the instantaneous center.

The usual remedy, which originated at GM about 1960, is to tilt the upper control-arm pivot axis up at the front so as to move the instantaneous center closer to the neutral line (Fig. 7-20). Mercedes-Benz has taken the same principle and added a further refinement. Mercedes-Benz has provided the ideal geometry by offsetting the upper and lower control arms against each other (in the longitudinal plane) so as to increase the anti-dive effect in direct proportion to the weight transfer.

Springs for Transverse A-Arms

Transverse A-arms can be combined with any sort of spring. In fact, some types of springs can be arranged to work as part of the locating linkage. Coil springs are the most common in modern SLA designs. As a rule, the coil spring stands on the lower A-arm and runs into an abutment below the upper A-arm. But the spring can also be arranged to stand on the upper A-arm, in which case it abuts into a spring tower that is usually part of the inner wheel housing.

The latter offers an open area facing the hub, as would be necessary for front-wheel drive. In both cases, the springs work in compression, and spring length varies less than actual wheel travel because the springs are mounted some distance inboard of the wheels and, therefore, have a shorter pivot arm. The importance of the pivot arm length comes into play in selecting spring rate (just as for shock absorbers).

The springs do not necessarily stand vertically, but are often inclined inwards at the top to conform more closely to a perpendicular position relative to the control arm that carries its base. The springs may also be mounted inboard of the upper control-arm pivot shaft, which then carries a rocker arm working against the top of the spring, whose base is part of the chassis frame. Depending on rocker arm length, this type of design permits a shortening of spring deflection, use of shorter or smaller-diameter but stiffer springs, and carries the entire spring-as-sprung weight.

Another popular type of spring for transverse A-arms is the torsion bar. It is usually arranged in the longitudinal plane, forming a physical extension of the lower control-arm pivot shaft, with its rear end anchored in a frame cross-member or part of the cowl structure. One design exists (Dodge Aspen and Plymouth Volare) where transverse torsion bars are linked with transverse A-arms. To make this combination possible, the Chrysler engineers came up with a crank and lever linkage to translate the up-and-down motion of the lower control arm into a twisting movement of the torsion bar.

Leaf springs are not much used nowadays in independent front suspension systems, but once were more popular than coil springs for transverse A-arm designs. A transverse leaf spring with its outer ends attached to the lower (or upper) control arms gives ample wheel travel, progressive spring rates and low unsprung weight. Transverse leaf springs have been used as locating members, replacing the upper or lower control arm. This works fine as long as the other A-arm has a sufficiently wide base to accommodate fore-and-aft stress loads without putting excessive bending loads into the leaf spring.

If a transverse leaf spring is used as part of the locating linkage, it is the position of the spring clamps that dictates the length of the control arm it forms—modified by any abutments above and below, which it may contact during deflection. Such abutments may be used to advantage in providing variable geometry and variable-rate springing. But leaf springs impose a severe limit on wheel travel if used as upper control arms (where short length is usually wanted). As lower control arms, the leaf spring is more adaptable. What's more, its mass is then located at a lower level, which assures a lower center of gravity—not an insignificant factor since a multi-leaf transverse leaf spring can be a heavy assembly.

Four parallel transverse leaf springs arranged at two levels can be used to form a linkage that corresponds to a system with equal-length A-arms. The separation between the springs at each level would substitute for the triangulation of the A-arms, and the outer ends of each pair of springs would be joined to the upper or lower king-pin bearing. Designs using this principle have been used for both front (Alvis) and rear (Cottin-Desgouttes) driving wheels and remain historical curiosities. But medium-size cars with SLA suspension and a lower-transverse leaf spring have been produced until quite recent times, notably by Peugeot (Types 203 and 403); a small car (Fiat 126 using a similar layout is still in production (Fig. 7-21).

Rubber springs, air springs, and suspension struts containing a combination of pneumatic and hydraulic chambers can be arranged to fit almost anywhere coil springs will fit. Their particular characteristics will be discussed in a later chapter. As far as the wheel-locating duties are concerned, they conform to the same rules that govern transverse A-arm suspension systems with metallic springs.

Two systems of independent front suspension, each offering great differences in geometry from those described so far, have had some importance historically, but are no longer used on any production car. They are (1) sliding pillar and (2) Dubonnet suspension. Brief descriptions

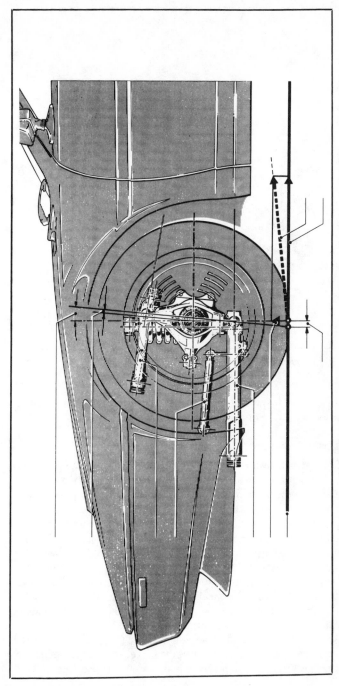

Fig. 7-20. Front suspension of BMW M-1 has strong backwards tilt on the upper control arm to counteract nose-dive. Lower A-arm is horizontal and linked to a stabilizer bar.

107

Fig. 7-21. Transverse leaf spring on Fiat 126 serves as lower control arm and elastic medium at the same time.

and a rundown of their pros and cons are included in the chapters dealing with suspension system evolution.

MacPherson Suspension

The system known as MacPherson suspension has become extremely popular for light-and-medium-size cars in the last 25 years (Fig. 7-22). It can be described as a compromise between the sliding pillar principle and that of transverse A-arms. Physically it has some parts in common with the latter. The lower control arm is similar. But the upper control arm has been replaced by a tall spring leg whose upper end is anchored in the body structure (wheel housing or cowl extension)(Fig. 7-23). This spring leg stands on a base that extends from the wheel hub, replacing the upper ball joint, and telescopes to permit wheel travel. The inner member undergoes partial rotation to permit the wheel to be steered. The spring leg is usually equipped with a coil spring surrounding it, near the top end, and contains a hydraulic damper as an integral part.

At a superficial glance, it looks as if the spring leg, tilted inwards and backwards from its base, actually forms the swivel axis and determines the caster angle (Fig. 7-24). In practice, however, that is not the case. The

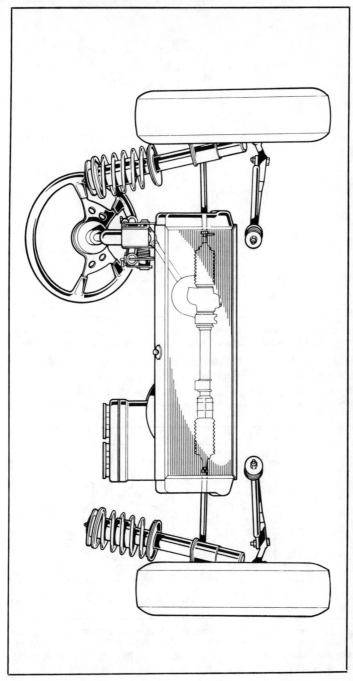

Fig. 7-22. Simplicity itself is the MacPherson front end of the Fiat X-1/9, with inclined spring legs and a lower lateral arm with drag strut.

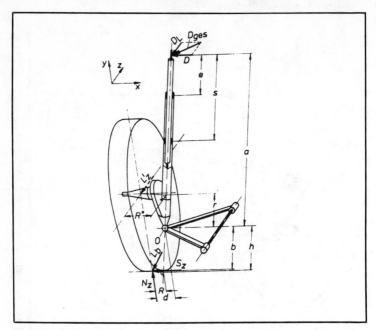

Fig. 7-23. Model of the Mac Pherson suspension shows the wide spacing of the attachment points, which makes the system very sensitive to wheel-balance.

swivel axis is formed by the straight line between the lower ball joint—that is, the hub attachment for the lower control arm—and the top anchorage point of the spring leg (Fig. 7-25). The location of the spring leg does not enter into the picture as far as the swivel axis is concerned.

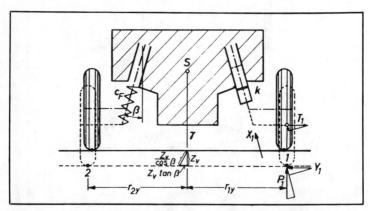

Fig. 7-24. Track variations in Mac Pherson suspensions depend on the inclination of the spring legs and the available deflection travel. Camber changes are dictated by lower-control-arm length and pivot-axis height travel. Camber changes are dictated by lower-control-arm length and pivot-axis height.

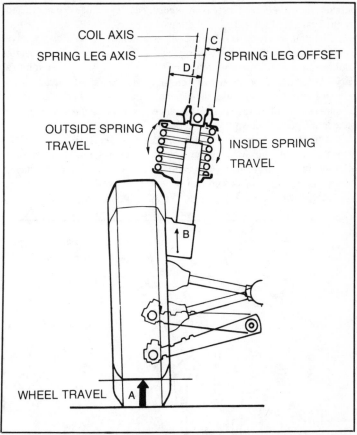

COIL AXIS
SPRING LEG AXIS
C
D
SPRING LEG OFFSET
OUTSIDE SPRING TRAVEL
INSIDE SPRING TRAVEL
B
WHEEL TRAVEL
A

Fig. 7-25. By careful interplay of control arm and spring leg geometry, Mitsubishi obtains a constant zero-camber angle on the Mirage (Colt) with MacPherson suspension.

Relative to the spring leg angle, the swivel axis can be more upright or made to lean more. It can even be arranged to coincide, but such cases are rare. The rearward tilt of the spring leg does not, in itself, determine the caster angle, which is fixed by the same points that dictate the swivel axis inclination. If the spring leg is tilted backwards so that its center line coincides with the lower ball joint, the caster angle corresponds exactly to the backwards tilt of the spring leg. On many cars, this is the case.

But geometrical considerations do not always make it practical to tilt the spring leg exactly the same amount that is needed to give the offset-base MacPherson spring leg, which originated on the BMW 2500 (Bavaria) in 1969. The BMW spring leg was tilted backwards at 14.5 degrees giving a caster trail of two inches; caster angle was 9.5 degrees. On the BMW, the steering axis inclination tends to give increased positive camber on the

111

outside front wheel with increasing steering angles. This is counteracted by the rearwards tilt of the spring leg.

Simultaneously the inside front wheel goes into increasingly negative camber angles with higher steering angles. This arrangement provides additional camber thrust for both front wheels as they are steered into a curve. As a result, the amount of steering input needed for a given lateral acceleration gets lower as the car's speed rises. The car which has little or no understeer at low speed gradually develops more understeer at higher speeds for greater down-the-road stability.

A further refinement was added on the 7-series BMW in 1977. It is known as the dual-link MacPherson system. Two links, one lateral and one semi-trailing, each having its own ball-joint connection with the spring leg base, assure that the design can be set to vary the scrub radius in accordance with the steering angle (Fig. 7-26) (small for straight-ahead

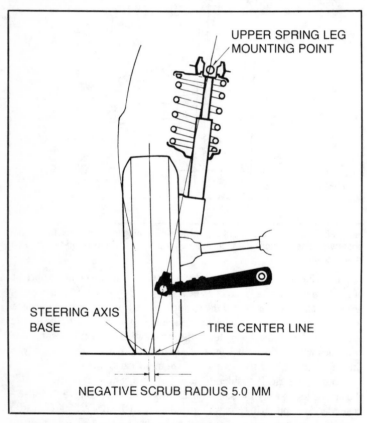

Fig. 7-26. MacPherson spring legs lend themselves just as easily to center-point steering or positive scrub radius as this negative-scrub-radius setup on the Mitsubishi Mirage (Colt).

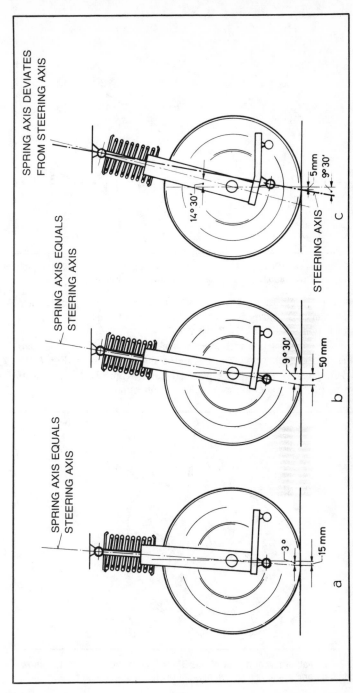

Fig. 7-27. BMW engineers have developed MacPherson suspension systems with double-offset spring legs.

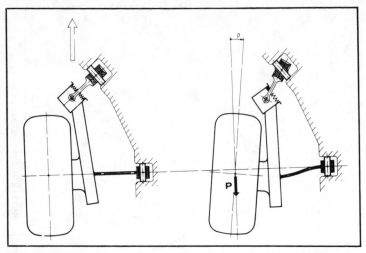

Fig. 7-28. The Mac Pherson design for the BMW 7-series has a moderately high roll center and a stabilizer bar linked to the lower control arm.

driving and progressively greater as the curves get sharper). It also provides an increased measure of anti-dive effect without affecting the roll center height. (See Figs. 7-27 and 7-30).

Roll center location with MacPherson suspension systems depends on the angle of the lower control arm and spring leg inclination (Fig. 7-31). The spring leg has effectively replaced the upper control arm, and it is its inwards tilt that guides the wheel during deflections.

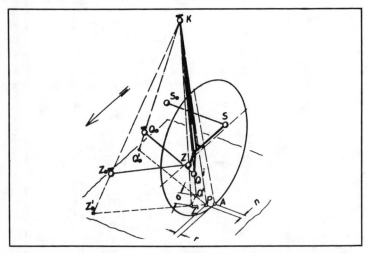

Fig. 7-29. Mathematical model of the BMW double-offset spring leg shows additional complications over the straightforward MacPherson type.

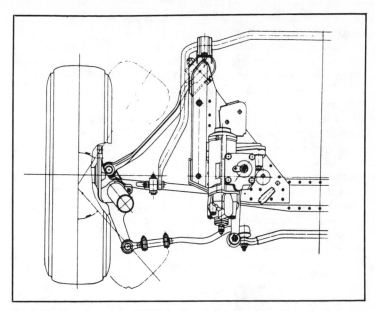

Fig. 7-30. Plan view of the BMW double-offset spring-leg suspensions shows details of steering linkage and full-lock positions of wheels.

To calculate roll center height, trace the usual straight line extending from the lower control arm on one side towards the opposite side of the vehicle. From the upper mounting of the spring leg, trace a line that runs perpendicularly to the spring axis. Where this line intersects the other one, you have the instantaneous center. From here on, it's the same as for transverse A-arms. Draw the line from the instantaneous center to the

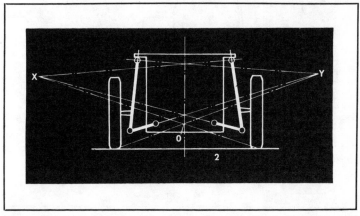

Fig. 7-31. Roll-center height with MacPherson suspension is derived from the angle of the spring legs and the angle of the lower control arms.

115

center of the tire footprint, and the roll center is situated where this line intersects the car's vertical center line (Fig. 7-32).

How does this work in practical examples? A horizontal lower control arm and a 20-degrees-from-vertical spring leg put the instantaneous center just outside the opposite front wheel and give a roll center slightly above ground level. Moving the spring leg closer to upright without changing the lower control-arm pivot axis will displace the instantaneous center farther away from the car and bring the roll center closer to ground level. A greater inclination of the spring leg will have the opposite effect—raising the roll center, but only to a very small extent.

Big changes in roll center height cannot be obtained by tilting the spring leg. It must be done by raising or lowering the lower control-arm pivot shaft. Raising it also raises the roll center. But how does that affect wheel camber? If the wheel has a static zero-camber in jounce, which is desirable. But it will also put the wheel into progressively severe position camber in rebound, which is undesirable in itself and simultaneously

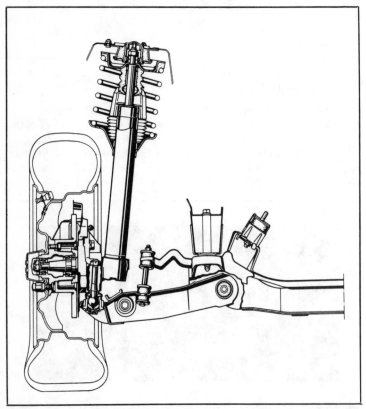

Fig. 7-32. MacPherson suspension on Opel Rekord gives a roll center about 2-1/2 inches above ground level.

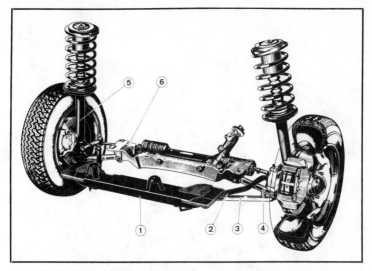

Fig. 7-33. One of the largest and heaviest cars to use MacPherson suspension is the Puegeot 604. Others are Ford Fairmont and Opel Senator.

narrows the track, which ought to be avoided. Consequently, there are strict practical limits to how much the engineers can juggle roll center heights in MacPherson front suspension systems.

As we have seen in the case of transverse A-arms, under dynamic conditions the roll center is displaced laterally as well as up and down for MacPherson suspension. But the displacements can be of a greater magnitude in MacPherson systems (see Fig. 7-33).

With MacPherson suspension, the roll center is displaced inwards on curves (as a result of body roll). It can move beyond the inside wheel in extreme cases, which makes it very important to restrict body roll. That means the roll center is divorced from the car's vertical center line under certain conditions. It can be helpful in restricting body roll, as long as the roll center is kept at relatively low levels, but if it rises to a higher level while being displaced laterally, it can aggravate the roll angle.

Adaptability of MacPherson Suspension

With regard to unsprung weight, MacPherson designs are more or less at parity with transverse A-arm systems. The spring leg may be heavier than an upper control arm, but its upper part with both the coil spring and shock absorber constitutes sprung weight.

MacPherson spring legs can also be adapted to other types of spring. Porsche developed a pneumatic spring with hydraulic damping for the 1969 model 911, fitting all the elastic and damping elements neatly inside the spring leg assembly. Its operating principle is similar to that of the gas-filled shock absorber. It is also self-leveling, maintaining the front end

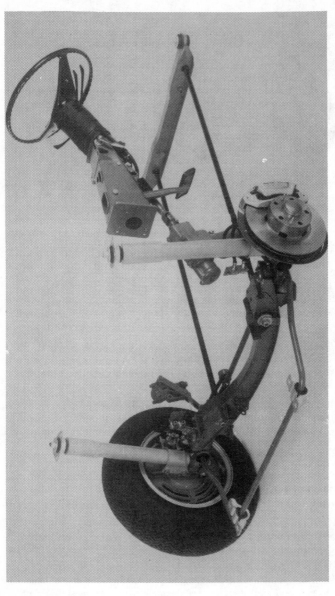

Fig. 7-34. MacPherson without springs? No, torsion bars are coupled to the lower control arms on the Fiat 130, eliminating the need for coil spring on the struts. Stabilizer bar forms drag strut and is an integral part of lower control arm.

at normal height regardless of load (with a built-in delay to allow wheel travel due to road surface irregularities).

Fiat produced an even simpler solution, removing the springs entirely from the legs and adding torsion bars to the lower control-arm pivot shafts on the Type 130 (1969). (Fig. 7-34) MacPherson suspension is naturally adaptable to a transverse leaf spring, installed with its end on the lower control arms; Fiat has used that idea for the non-driving rear wheels of the Type 128, but not in a production-type front suspension system.

The main attractions of the MacPherson type of suspension used to be low cost and ease of assembly, but these points have less validity today since the most satisfactory MacPherson designs tend to be quite elaborate assemblies with a high number of components.

Chassis development engineer Magnus Roland of Saab-Scania points out that there is often a ride problem with MacPherson suspension systems because of a stick/slip effect within the telescopic spring leg. This can best be overcome by higher precision in machining and assembly, more expensive bearings and seals, and special provision for lubrication of surfaces that operate with metal-to-metal contact.

In addition, says Roland, MacPherson suspensions tend to be very sensitive to wheel balance and tire uniformity. Other manufacturers, including some who use MacPherson suspension on their cars, confirm that this is true. Roland blames this on inherent dynamic weakness in the system due to the mounting points being so far apart. That very fact, on the other hand, is advantageous in distributing the shock loads from the suspension system into a widespread area of the body shell.

That can help towards lighter body and frame construction, which is a point of major importance in cars of all sizes. But among other arguments that can play against the use of MacPherson suspension is the streamlining trend which tends to flatten the hood and fenders and, thereby, cut down the available installation space for the spring legs. This restricts wheel travel and imposes stiff damper settings, as well as preventing the use of very soft springs.

Chapter 8
Rear Axle and Independent Rear Ends

Rear wheels are different from the front ones in several ways. First, they are not steered. That tends to simplify rear suspension design. It also makes a big difference to rear suspension design if the wheels are driven or not. Front-wheel drive opens the way for great simplification of the rear suspension.

As we have seen, it makes very little difference in independent front suspension design whether the wheels are driven or not (as long as the suspension system does not block the way for the drive shafts). But in rear suspension design, it makes a world of difference whether the rear wheels are driven or not. We shall soon see why and how.

While it is true that the American auto industry is switching to front wheel drive in a big way, the majority of its current products have rear axle drive. But the switch may never go across the board. And in Europe, there are car companies that remain basically opposed to front-wheel drive (such as BMW), so that we have reason to believe that cars with driven rear wheels will continue in production indefinitely.

Live Rear Axle and IRS

Just as rear suspension systems break down into two main categories for driven and non-driven wheels, the driven-wheel category is divided into two separate classes: rear axle suspensions and independent rear suspension systems.

We have briefly discussed the rear axle in Chapter Two, so we do not have to repeat its essential problems here. But what about its good points? According to a leading British engineer, Donald Bastow, it offers only two advantages. The first is that it is less expensive, particularly at the present time, Bastow wrote in 1951. The second is that we have considerable knowledge of its problems and their solutions or palliatives.

For the sake of completeness, we should also be reminded of the main characteristics of the rear axle in terms of suspension geometry:

- Both wheels remain at zero camber in cornering, but both change camber angle on single-wheel bumps;
- Track is constant;
- Horizontal compliance depends on detail design of the locating members;
- Unsprung weight is very high;
- Roll center is high (at spring anchorage height).

As you can see, control of wheel travel with a rear axle is the main problem. And without accurate wheel guidance, handling precision, roadholding and ride comfort are helplessly compromised.

The simplest rear axle suspension system, called *Hotchkiss drive,* consists of two semi-elliptic leaf springs mounted longitudinally (Fig. 8-1). Largely because of its inability to cope with torque reactions in the live axle, (Fig. 8-2) Hotchkiss drive is on the way out. One can only register amazement that it has held out in production as long as it has. It is still used on such cars as the Ford Pinto and Escort, Ford Granada and Mercury Monarch; Lincoln Versailles; the entire Plymouth, Dodge and Chrysler range except the front-wheel drive models; Toyota Corolla, plus the Carina and Corona station wagons; Datsun 210 and 200 SX; Morris Marina, MG Midget and MGB.

Springs must be long to assure proper ride comfort, but long springs are the least suitable for wheel-locating duties. Some makers, notably Chrysler, try to combat this problem asymmetrical spring design and axle mounting. The axle is moved forward from the spring center and the front

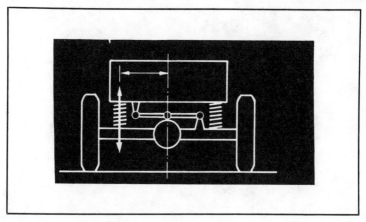

Fig. 8-1. With semi-elliptic leaf-spring suspension, the live axle gives a roll center at spring-anchorage height.

121

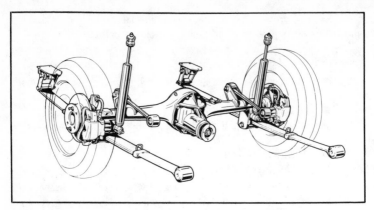

Fig. 8-2. Live axle of the Fiat 125 used twin-leaf springs of unusual length and radius arms on both sides to control torque reactions.

portion of the spring beefed up so it is better able to handle thrust loads. The long rear part is shackled and free to flex.

Placing the springs as close to the wheel hubs as possible gives a broader spring base and helps give roll stiffness. Wider leaves assist towards more positive lateral axle location. Judicious shock absorber installation can also bring benefits in this regard (Fig. 8-3). But nothing short of radius rods can assure control of roll-steer effects. Once you add radius rods, you no longer have Hotchkiss drive. And once you start building a proper wheel-locating linkage, there is little incentive to retain the semi-elliptic leaf spring, particularly in its position concentrating its mass around the axle clamps, which maximizes the unsprung weight.

Fig. 8-3. Initial version of the Fiat Dino Spider had a live axle carried on single-leaf springs, with trailing arms and dual shock absorbers.

For example, Jaguar's 2.4-liter sedan from 1956 went to a live rear axle located by a combination of cantilever springs and radius rods. But it would make equally good sense and bring further ride-and-handling improvements to adopt coil springs, torsion bars or non-metallic spring units (Fig. 8-4). Most rear-drive cars in current production have either gone to independent rear suspension or more advanced axle-location systems.

Axle Location Systems

The main objective of axle location systems is to restrict the axle's freedom of movement. It must be prevented from displacing itself sideways. It must be held in a near-pure transverse plane throughout and changes in wheelbase must be kept within close tolerances; but the suspension must permit up-and-down movements of each wheel alone, as well as make it possible for both wheels to rise or fall together.

The goings-on between the axle and the frame, or body structure, occur in all possible planes and combinations of planes. First, there is the *vertical force* due to the weight of the body and its payload. Secondly, there are the *longitudinal forces* generated by tractive effort and braking. Thirdly, there are *torque forces*, rotational or twisting, stemming from reactions to tractive effort and braking. Fourthly, there are *transverse loads*, placed on the axle and suspension system by centrifugal force, sidewinds and cambered roadways.

A linkage that provides accurate guidance for axle movement in the needed planes must also be devised so as to prevent axle movement in other planes without imposing connections that fight each other or interfere with the desired wheel travel as demanded by ride comfort considerations.

Torque Tube. One method to control longitudinal forces and axle-twisting torques is the *torque tube*. In its basic form, it consists of a tube enclosing the full length of the propeller shaft. The rear end of the torque

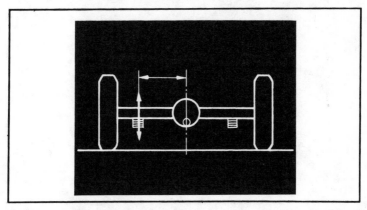

Fig. 8-4. Live axle with coil springs and track bar has its roll-center height fixed by the lateral-location linkage.

tube is bolted to the differential nose piece of the axle casing. The front end of the torque tube is pivoted in a ball-joint mounting or fork-type mounting at the output end of the transmission.

On more recent designs, such as the Fiat 124, the torque tube is just a short piece leading to a cross-member or other parts of the main chassis or body structure capable of taking up the driving thrust loads (Fig. 8-5). Torque tubes effectively resist the torque reactions that tend to spin the axle housing in the direction opposite of wheel rotation, but do little to alleviate the torque reactions that tend to spin the axle housing around its center—or more accurately, around the contact point of the pinion and ring gear. Therefore, the torque tube cannot do much to prevent wheel hop.

Torque tubes fail to provide adequate lateral location of the axle and are usually supplemented by diagonal stays (a form of semi-trailing arms) or radius rods. Torque tubes add substantially to the unsprung weight, especially if combined with the longitudinal, semi-elliptic leaf springs clamped to the axle casing at their middle. To partially offset this gain in unsprung weight, engineers have sought other spring types for torque tube drive systems, such as cantilever springs (Rolls-Royce), transverse leaf springs (Ford) and coil springs (Buick).

Four-Link. Today's typical axle location system has four links and vertical coil springs (Fig. 8-6). The lower pair of arms—running at or below axle center level—usually trails and takes up the tractive and braking thrust. These trailing arms are usually widely spaced for maximum resistance to rear wheel steering phenomena. On some cars

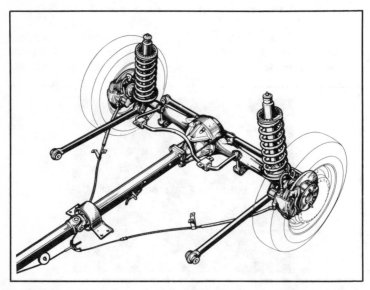

Fig. 8-5. Fiat's 124 set a new trend in 1966 with the revival of the short torque tube. Long radius arms are placed as closely as possible to the wheels.

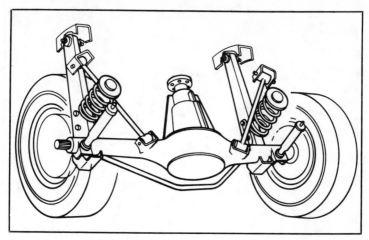

Fig. 8-6 Triumph's TR-7 of 1975 had a four-link system of axle location with coil springs mounted on the trailing arms.

they have slight semi-trailing (diagonal) angles, aimed from near the wheel hubs to an imaginary point somewhere on the car's center line, usually in the transmission area. In all cases, they tend to be long (equal to about half the track), giving a long radius for wheel travel so as to restrict wheelbase variations.

The upper pair of arms is usually much shorter, attached high up on the axle casing and widely splayed. Typically, they are anchored on the differential housing and, therefore, close together at their rear mounts, running diagonally at 45 degrees or so to more widely-spaced brackets on the chassis frame or body shell (Fig. 8-7). Their main task is to counteract torque reactions in both ways-axle rotation around (a) the pinion and (b) the ring gear. They also provide sideways location of the axle, admitting no greater lateral shift in the axle than the flexing of the rubber bushings in the anchorage points (Fig. 8-8).

Three-Link Plus Panhard Rod. Many variations of the four-link axle-location system exist. One is the three-link plus track-bar system, in which the upper pair of arms is replaced by a single trailing arm on the right side. It's a torque-reaction member and quite effective as such, but provides no lateral stability. The track bar is added for that purpose.

A track bar is often called a *Panhard rod*. It is a straight rod or bar running transversely across the chassis behind the axle. One end is attached to a bracket on the axle casing, and the other end to a bracket on the frame or body. Thus, it keeps the axle in a permanent lateral relationship to the sprung mass.

Permanent? Well, not exactly, for it dictates a slight sideways displacement of the entire axle since the track bar swings through an arc. The longer the track bar, the straighter the arc. But when the track bar is attached to the left frame side member and to a bracket on the right side of

125

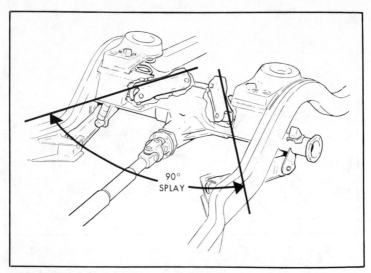

Fig. 8-7. Buick and other GM cars avoided the track bar by splaying the upper control arms at right angles to each other so as to take care of lateral location.

the axle casing, a few inches lower down, it is, nevertheless, unavoidable that the entire axle assembly should be pulled to the left in rebound while moving to the right (and possibly back again) during jounce deflection.

In such systems (three- and four-link), coil springs are most commonly used. They can be mounted vertically or slanted slightly forward or

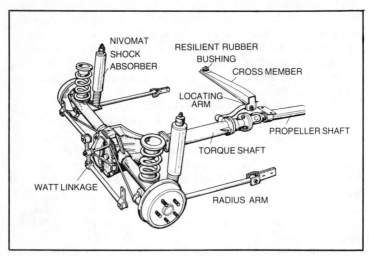

Fig. 8-8. Live axle of the Rover 3500 of 1976 was located by twin radius rods and a Watt linkage. A short torque tube was linked to a structural cross-member anchored in rubber bushings.

inboard, or both. Sometimes they stand on top of the axle casing, sometimes behind and sometimes in front. They play no part in axle location, and their position is dictated mostly by ride comfort and installation-package considerations.

Countless variations on the same theme exists with basically similar geometry. Alfa Romeo, for instance, uses a single T-shaped link (Fig. 8-9) in place of the splayed upper arms, giving more positive lateral location without giving up any of its ability to control torque reactions.

On some systems, two parallel radius rods take the place of the single trailing arm on each side. This system is usually employed when space problems exist, limiting the length of the arms. With short trailing arms, the axle housing would move in an arc giving relatively important wheelbase variations and involving partial rotation of the axle housing. With upper and lower radius rods, the partial rotation is eliminated.

Watt Linkage. Instead of the splayed upper control arms, or a single T-arm, some cars (notably Bristol) uses a *Watt linkage*. It is called that because the principle was first used to operate the valves on steam engines by James Watt. It can be considered as a dual track bar with a rocker arm in the middle. The rocker arm center is attached to a pivot point on the geometrical center of the axle housing. It is a straight rod which assumes a tilt angle, say two o'clock and eight o'clock, in the static position. The ends of the rocker arm are linked to the ends of semi-track bars attached to joints carried on the frame or body shell at either side. The Watt linkage permits unrestricted axle movements up and down, within the dimensions of the rocker arm which moves towards a vertical position in both jounce

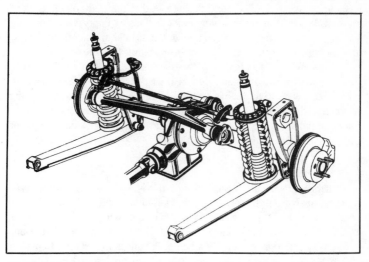

Fig. 8-9. Alfa Romeo's T-bar, used on the Giulia and its 1750 and 2000 successors, did double duty as a torque-reaction member and lateral-location link.

and rebound (as the distance increases between the axle center and the outer end joints of the semi-track bars). But the axle is effectively prevented from making any lateral displacement.

While coil springs remain the most common elastic element for such axle location systems, some designers have chosen to combine them with torsion bars—transverse or longitudinal—and even leaf springs have been used (Ferrari). To engage a longitudinal torsion bar, a simple lateral bell-crank is enough, and for a transverse torsion bar, the bell-crank can simply be turned to a trailing or leading position.

IRS Eliminates Torque Reaction

One is tempted to observe here that the more sophistication that is put into axle location systems, the greater their cost and complication. This point has no doubt been caught by a number of engineers who, as a result, decided to do away with the axle altogether and go to independent rear suspension.

The most important advantage of independent rear suspension systems is the complete elimination of the torque reaction effects. With the final drive unit separated from the axle and mounted rigidly in the chassis frame or body shell, there can be no torque reaction in the driving shafts, as all torque reactions are absorbed by the frame or body structure.

The wheels are driven by open half-shafts, often splined to allow variations in length and equipped with one or two universal joints to permit wheel travel. These shafts can be regarded as *transverse propeller shafts*. The suspension linkages, naturally, tend to resemble independent front suspension systems. The big difference is that the rear wheels are not steered. Whether they are driven or not is of lesser importance since the locating members must be strong enough to handle braking thrust loads, and they can be several times greater in magnitude than the driving thrust.

But—and here's a big but—if the rear wheels are not driven, there is less to gain by going to independent rear suspension. An I-beam rear axle on a front-wheel drive car has no torque reactions and offers interesting characteristics in terms of roll stiffness, camber changes and choice of elastic element, while not adding significantly to unsprung weight or posing space-allocation problems.

Live Axle Drawbacks

In the following discussion, we are, therefore, concerned with driven wheels. Apart from the torque-reaction problems, the live axle has many other drawbacks. It is the cause of many different disturbances transmitted to the vehicle structure from (a) the road wheels and (b) the drive train.

When one rear wheel hits a bump, the shock forces tend to push it upwards and backwards. Because both wheels are carried on the same axle, a downward and forward impetus is transferred to the other rear wheel. This impetus tends to detract from the ride comfort and handling precision of the car, and the axle's behavior is a major cause of noise, vibration and harshness. As we have seen in the case of I-beam front axles, a single-wheel bump causes camber changes in both front wheels. With rear axles, the situation is the same, but the disturbance is magnified by the greater unsprung weight.

The live axle is still preferred by many of the world's leading auto makers and is still a feature of many truly good cars—Chevrolet Camaro and Pontiac Firebird, Buick Century and Oldsmobile Cutlass, Fiat 131 and 132, Volvo 244 and 264, Ford Mustang and Mercury Capri, Mazda RX-7 and 626, Opel Rekord and Commodore (the Peugeot 504 is available with independent rear suspension in its GL and TI versions, though the baseline model has a live axle).

"Good enough" has always been the worst enemy of "The Best" and no doubt many car makers justify their choice of a live rear axle on those grounds. In addition, they are afraid of the unknown, especially in the area of cost. They have been led to conclude that the independent rear suspension is not worth the extra cost. Let me insert a dissenting voice.

It can, in fact, be argued that the advantages of independent rear suspension are worth more than their cost. With an unlimited budget, no engineer can achieve all the same benefits as long as he must use a rear axle. It is certain that it would cost more to optimize a rear-axle-driven car in terms of ride and handling, noise and vibration, traction, weight saving and space utilization, than to quite simple design and develop an independent rear suspension system for it.

The economics of making cars do not stop there, however. There is also the cost of making changes in the manufacturing setup. Many companies have huge investments in making axles and have a natural desire to protect them. But even axle-making machinery wears out and must be replaced. And, when the time comes, it could be replaced by tooling and equipment for the components used in independent rear suspension systems. Consequently, the tooling-change argument is not one of principle, but one of timing. It will merely cause a delay in the death of the live axle.

Benefits of IRS

A brief review of the benefits to be gained from independent rear suspension will make it clear that its adoption is the best means for makers of rear-wheel-drive cars to fight back against the front-wheel drive revolution.

Improved Ride Comfort. A high unsprung-to-unsprung weight ratio is beneficial to ride comfort. Compared with a typical live-axle installation, independent rear suspension lowers the unsprung weight to about one-third. Since axle weights do not vary greatly with car size, it is clear that the axle and its locating system constitutes an increasing proportion of the total weight, the smaller and lighter the car is. Since we live in an age of down-sizing, retention of the live axle would entail accepting a serious drop in sprung-to-unsprung weight ratios, with resultant deterioration in ride comfort.

With independent rear suspension, spring rates can be selected that will give near-optimum ride comfort on a greater variety of road surfaces and will tolerate greater variations in payload without severe loss of ride comfort. This is a point of vital importance for small cars, since the payload represents a greater proportion of the gross vehicle weight, the lower the car's curb weight.

Improved handling. A jolt on one rear wheel carried on a live axle will produce a side force on the other rear wheel. And body roll causes spring deflections in opposite directions left and right. Both conditions tend to give rise to rear wheel steering effects that are difficult to control. With independent rear suspension, single-wheel bumps have no effect on the other wheel, and roll-steer effects can be carefully controlled.

The reduction in unsprung weight also assists towards better handling, as a wheel with a smaller unsprung mass attached to it will find it easier to adhere to the roadway than a wheel that's part of a live axle assembly. Wheel geometry can be arranged to answer the chassis engineer's wishes within close tolerances, whereas the live axle imposes strict limits in this regard.

With more accurate wheel guidance, particularly in terms of camber changes, the engineer can fine-tune the suspension design for optimum roadholding. Independent rear suspension also offers the engineer the opportunity of placing the roll center at the height he wants it for the best interplay with the wheel locating members and spring rates.

Since torque reactions are eliminated from the rear suspension, there is no lift force on the right rear wheel during acceleration. That means improved traction (reduced wheelspin), and that is a special safety factor under low-friction driving conditions (rain, ice, snow, slush and mud).

Better space utilization. With independent rear suspension, adequate wheel travel can be provided within smaller dimensions than when a whole rear axle is displaced in jounce. Since the final-drive unit is bolted to the frame or body shell, there is a chance to gain considerable space along the floor. The propeller shaft can be lowered, lowering the center of gravity, which also contributes to improved handling.

The rear seat can be moved backwards, adding rear leg-room. Fuel tank and spare wheel can be repositioned for the most practical trunk space. Moving the fuel tank forward, adjacent to the rear seat back, helps towards evening out the weight distribution and lowers the polar moment of inertia, while it also gives the tank better protection in a rear-end collision.

Better brake cooling. Independent rear suspension offers the engineer an opportunity to remove the brake drums or discs from the wheel, to be carried near the inboard end of the drive shafts. In this location, the brakes can get the benefit of an open or ducted air flow greater than what's possible with wheel-mounted brakes. That means the brakes will run cooler, which gives longer lining life and reduces the risk of fading. Inboard brakes are carried as sprung instead of unsprung weight, which further enhances the ride-and-handling picture.

Thus we see that the advantages of independent rear suspension overlap in several ways. While its benefits in terms of active and passive safety depend greatly upon detail design, it is certain that cars with independent rear suspension can offer both a greater ability to make evasive maneuvers or stop better, thereby minimizing the risk of accident involvement, and simultaneously permitting the designers to lay out the chassis with higher regard for the safety of such sensitive items as fuel tanks.

Swing Axles

The simplest form of independent rear suspension (for driving wheels) is the swing axle system (Fig. 8-10). Though it is still in use, notably on the Triumph Spitfire, it is generally considered to give rise to more problems than it solves and, while swing axles were once the most common type of independent rear suspension, most car makers who used them have abandoned them in favor of more advanced systems.

A typical swing axle design consists of a final drive unit carried integrally with the frame or body shell, having a universal joint on each side, with open drive shafts to each wheel hub (Fig. 8-11). Since there is no universal joint at the hub end, the axles make pendulum-type movements around their inner joints when the wheels hit obstacles in the roadway. If outer joints are fitted, you no longer have a swing axle system.

This pendulum-action is at once an essential characteristic of swing axles and the source of one of their main drawbacks. The wheels undergo very important camber changes—towards negative in jounce and positive in rebound. That does not sound so bad, for theoretically, that would put the outside wheel at negative camber when cornering. But, for other reasons that I will explain in a minute, that is rarely so in practice.

First, it must be realized that the pendulum action does not occur in a straight up-and-down line. On all but the least powerful economy cars, the swing axles need the support of radius arms to take the driving thrust, and the wheel-hub end of a radius arm naturally describes an arc when the wheel is deflected.

As a result, the wheel is pulled forward towards the extremities of its travel. Since the guidance of the wheel is determined by the location of its two pivot points (the forward anchorage of the radius arm and the universal joint on the drive shaft), the wheel is steered on a diagonal axis. That means it also changes toe-in on deflections. If the wheels have a zero toe-in static setting, they will move progressively towards greater toe-in angles in both jounce and rebound.

Jacking Effect. Swing axles, clearly, are subject to very substantial rear-end steering phenomena due to their inherent geometry. These may

Fig. 8-10. Swing axles of the Volkswagen Beetle were enclosed and pivoted from universal joints on either side of the differential. Outer ends were guided by trailing arms that operated short-transverse torsion bars.

not be important on single-wheel bumps when the car is traveling straight down the road. But roll-steer effects when cornering can be severe, and worse still is something that is called a *jacking effect*. This jacking effect is the result of the swing-axle geometry providing a very high roll center. As we have seen before, a high roll center is an advantage for cars with rear axles. But due to the problematical geometry of swing axles, it becomes a problem in that combination.

To explain the jacking effect, it is best to play the game of trying to predict what will happen, and then compare it with what actually happens. As the swing-axle car goes into a cornering situation, weight transfer at the rear will be high because the roll center is high. That means unloading the inside rear wheel and placing more weight on the outside one. Theoretically, the outside wheel will move into jounce, with increasing negative camber, and develop greater cornering force (resistance to side forces). The inside wheel will be free to drop into rebound and, since it is lightly loaded, its positive camber will not matter at all. Right? Wrong.

The outside rear wheel never gets a chance to settle into a negative camber attitude. Why not? There is an interesting interaction between various design elements and the physical laws affecting a car going through a curve.

First, there is a certain forward weight transfer due to the car's yaw velocity (because the front tires absorb more energy when steered away from the straight path, and this produces a braking effect on the car). The result is that some load is taken off both rear wheels and added to the front ones.

The height of the rear roll center then comes into play. To find it, draw a straight line from the center of the tire footprint to the drive-shaft universal joints-and beyond. Where the line crosses the car's vertical center line, above or in the upper part of the final drive unit, is the roll center. That can be 15 to 20 inches above ground level.

This gives high roll stiffness and strong weight transfer. Whatever roll angle is reached, it cannot produce negative camber in the outside wheel because the camber angle on the inside wheel prevents the final-drive unit from coming down. Once the outside wheel has been moved into a positive camber situation, the jacking effect sets in. Each transient load change, occasioned by road surface unevenness and variations in yaw velocity, serves only to tilt the rear wheels further into rebound (and positive camber), pushing the final drive unit and the drive-shaft universal joints higher and higher. It is as if the two drive shafts link arms to jack up the tail end of the car (which also puts the car into a nose-down attitude).

Because sidebite is lost as the outside rear wheel tilts more and more, the wheel simultaneously begins to steer at a tangent of increasing angle from the arc of the curve and the car becomes unstable. The driver usually senses this as a growing oversteer, making it necessary to unwind the steering wheel. Increasing toe-in on the outside wheel is inadequate to counteract the camber-steering effect and the inside rear wheel, because it is relatively unloaded, cannot do much to help keep the car on its intended course.

When side forces get strong enough, the rear wheels will lose side-bite and begin to skid. For the driver, this represents rear-end breakaway. It is difficult to correct because it is capricious, again due to swing-axle geometry. If unchecked, it will lead to a spinout.

Springs and shock absorbers with high rebound resistance are ineffective in cortrolling the defects of swing-axle geometry. Auxiliary springs (camber compensators) can alleviate the problem, but cannot alter its nature.

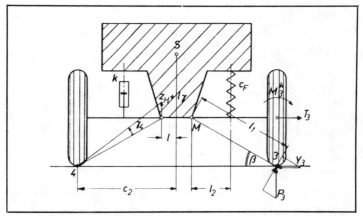

Fig. 8-11. Two-jointed swing axles with separate pivot points have a roll center located at the height of the universal joints.

Any type of spring can be used with swing axles. Coil springs can be mounted on the radius arms, for instance, and a transverse leaf spring can be fitted above or below the differential casing. Torsion bars? Mounted transversely, they can be combined conveniently with the radius-arm pivot shaft, and mounted longitudinally, they can be operated by lateral crank levers from the wheel hubs. pneumatic and oleo-pneumatic spring units can be installed anywhere coil springs can fit.

Swing-Axle Variations

Mercedes-Benz developed a variety of swing-axle systems with notable improvements. They included auxiliary springs (to fight excessive positive camber), sometimes using torsion bars, coils or hydraulic devices.

But the major development was the *low-pivot-point swing axle*, (Figs. 8-12 and 8-13) which effectively altered the geometry in respect of both pendulum-arm length and roll center height. To make this possible, the final-drive unit was divorced from the body structure and carried on a journal which was suspended by a laminated rubber cushion resting on a cross-member. A downward extension on the final-drive casing provided a bracket for a longitudinal pivot shaft. The left drive shaft was bolted to the differential casing which was free to rotate enough to permit wheel travel. Only one universal joint was used, positioned inside the casing, aligned with the pivot shaft.

The right drive shaft was enclosed in a similar tube whose inner section was open, but fitted with a carrier arm that was anchored on the pivot shaft. At the outer ends, the drive-shaft casings were held by radius arms that carried coil springs. The low-pivot point became the roll center,

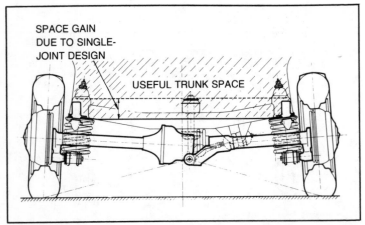

Fig. 8-12. With the Mercedes-Benz low-pivot-point swing axles, the roll center is lowered to the level of the single joint. Final-drive casing is integral with left-side swing-axle housing.

Fig. 8-13. Low-pivot-point swing axles of the 220 SE of 1961 had a compensating coil spring to control camber and a single post, rubber-mounted in the body shell, for lateral location.

located at a plausible height of about 7.5 inches. Since the single-pivot point replaced both the universal joints, it maximized drive shaft length and, thereby, gave the minimum change in camber angles for a given amount of wheel travel. Though this system was developed to a high degree of perfection, it was eventually replaced by designs based on principles offering the engineers greater geometrical freedom.

The reasons swing axles have survived as long as they have lie mainly in the basic advantages they share with other independent rear suspension systems—absence of torque reactions, improved traction, space-utilization gains and increased ride comfort—but the handling and stability problems of swing axles have barred their use from modern cars.

Lateral A-Arms

Independent rear suspension systems using lateral A-arms are derived from front suspension systems. Because we are talking about driven rear wheels, they form a direct counterpart to front-wheel drive SLA systems, minus the steering swivels.

They have the same advantages, such as very precise wheel guidance, low unsprung weight and wide freedom to choose roll center height. They can be combined with any type of elastic element. Transverse leaf springs can be installed between either pair of control arms. Coil springs, naturally, must be placed so that they do not interfere with the drive shafts. Torsion bars are usually arranged longitudinally, coupled to the lower control-arm pivot shafts. The drive shafts (often splined to permit length variations) are double-jointed, with a pair next to the diffe-

135

rential and one at each wheel hub. Camber and toe-in changes are determined entirely by the control arms.

A variation which has no parallel in front-wheel drive systems in the SLA type of rear suspension in which the drive shafts are used as upper control arms. In such designs, the shafts are beefed up and unsplined, of course. The joints are formed so that they can accept axial loads as well as torque. The lower control arm must be splayed so as to be able to take up the driving thrust on its own. Elimination of separate upper control arm means reduced unsprung weight and lower cost. It does not necessarily mean restrictions on roll center height, or impose undesirable camber changes in roll. Horizontal compliance is usually provided by rubber bushings.

The Weissach Axle

What Porsche calls the *Weissach axle* (Figs. 8-14 and 8-16) is not an axle at all, but an independent rear suspension system named after the place it was developed—the company's proving grounds at Weissach. It is a variation within the lateral A-arm family. To start at the wheel, the hub carrier has a triangulated lower control arm and a single I-bar upper control arm. Coil springs stand on platforms extending backwards from the hub, with concentric shock absorbers. The springs lean forward at an angle of about five degrees and inward at about 24 degrees.

What's so special in the Weissach axle is the lower control arm. Its lateral member is a slightly cranked rod, which changes angularity during deflection and moves the hub out of the strict vertical plane.

The other member is a diagonal drag strut with a novel type of articulation. This articulation is located near the front anchorage, where the forward end of the main link is formed into a cylinder aimed diagonally towards the midpoint of the car. But it is not an integral part of the main link. It is pivoted and contained in a holder with a rubber bushing.

Inside the cylinder is a short rod whose forward end pivots in a narrow bearing, with a thick-section rubber bushing, lodged in a bracket on the body shell. The articulation is made possible by the flexibility of the rubber bushings, and movement is restricted to very small changes in drag-strut length.

The big idea about the articulation was to prevent toe-out in rebound-during deceleration, for instance, and especially on curves. As forward weight transfer sets in, the wheels move into rebound. The crooked rods swing backwards at their outer ends, moving the wheel hubs backwards. This puts the articulation into play, stretching the drag strut. The wheel hub is steered inwards, making the wheel toe-in.

The effect is to provide a kinematic effect in the suspension system that balances the forces trying to make the wheel toe out. As a result, the Weissach axle is free of rear-end steering tendencies to a greater degree than other independent rear suspension systems.

The drag struts work as compression struts during acceleration, and because the articulation is formed link a diagonal pivot axis extending from

Fig. 8-14. Rear suspension on Porsche 928 is called Weissach-axle and belongs in the semi-trailing arm family, with a pivot axis about 60 degrees from transverse.

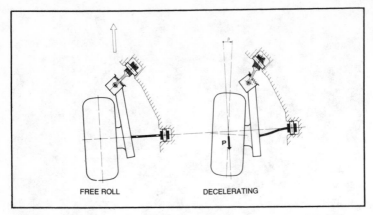

FREE ROLL   DECELERATING

Fig. 8-15. Flexibility in Weissach-axle links and mountings is designed to put the wheel into a toe-in position in rebound (when other semi-trailing arm systems tend to toe out).

another pivot axis in a flexible assembly of compact dimensions, it is fully capable of transmitting the drive thrust to the body structure. Stress loads in compression keep the strut at its minimum length, with zero toe-in, and increasingly negative camber as rewards weight transfer moves the wheels into jounce positions.

The drag strut is angled downwards from the pivot to the hub with a static setting of about 20 degrees from horizontal. This geometry provides a slight horizontal compliance on initial jounce travel, assisted by the rubber bushings.

On the Weissach axle design used on the Porsche 928, the wheels never go into positive camber, but remain upright to the end of rebound travel. Though this design uses coil springs, one can easily visualize

Fig. 8-16. Weissach-axle, seen from the front and minus drive shafts, shows non-aligned pivot points of the semi-trailing arms and pronounced inboard inclination of the coil spring and shock-absorber units.

138

torsion bar systems used with a Weissach axle (transverse, worked by cranks from the wheel hubs for longitudinal, extending from the upper control-arm pivot shaft). Non-metallic spring systems are also adaptable to the Weissach axle.

Trailing Arms

A single trailing arm per wheel can provide a rear suspension system of considerable merit for some types of car. The trailing arm carries the wheel hub, and the wheels are driven by double-jointed shafts. Trailing arms are most conveniently combined with transverse torsion bars, or with vertical coil springs standing near the aft end of the trailing arms, though other elastic elements can also be accommodated without major difficulty.

The rear wheel geometry permits no variations in track. Camber changes correspond to the body roll angle, and the roll center is located at ground level. Wheelbase changes vary with the length of the trailing arms (Fig. 8-17), which are usually carried on rubber bushings to give horizontal compliance.

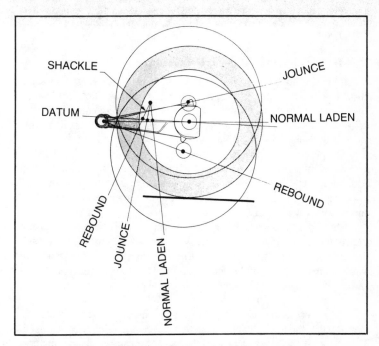

Fig. 8-17. Swinging arc of trailing arms can be utilized to add to the horizontal compliance of the rubber bushings at the pivot points. Wheelbase variations are greatest with short trailing arms, minimal with long trailing arms.

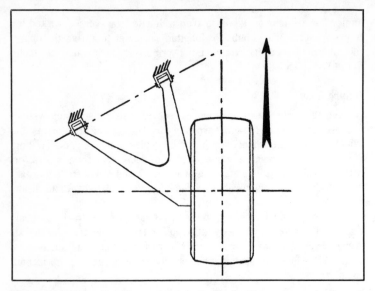

Fig. 8-18. With semi-trailing arms, the wheels pivot on a diagonal axis and, therefore, become subject to camber and toe-in changes. Roll centers can be raised or lowered, depending on pivot-axis height and the angle of the semi-trailing arm.

Semi-Trailing Arms

It is the trailing arms, of course, that take up both driving and braking thrust forces and locate the wheels laterally, as well. Trailing arms assure low unsprung weight and relatively low cost. But the geometrical limitations of the system have led the car makers away from trailing arms towards a compromise solution known as *semi-trailing arms*.

The difference lies mainly in the angle of the pivot axis (Fig. 8-18). It runs laterally across the car when trailing arms are used. Semi-trailing arms have a pivot axis bent back at the inboard end, at angles from about 10 to 40 degrees. This angle determines the toe-in, track and camber changes during wheel deflection.

The wider the angle, the larger the changes. The basic pattern is that jounce deflections give increasingly negative camber, while in rebound, the camber changes to positive. Correspondingly, the wheel tends towards increased toe-in in jounce, but towards toe-out in rebound. Track variations are minor, but increase with the angle of the pivot axis (from transverse).

By lowering the inboard pivot points, the pivot axes can also be moved out of the horizontal plane so that roll center height can be raised. This slope of the pivot shaft also modified the camber-change picture, though without altering the basic characteristic imposed by the semi-trailing angle.

Chapman Struts

Independent rear suspension systems characterized by their use of Chapman struts are basically rear-end adaptions of MacPherson-type front suspension designs (Fig. 8-19). They have two-jointed drive shafts, usually splined, so that the shafts are relieved of all stress loads except the driving torque. The spring leg can be made simpler than for a MacPherson strut, since it does not have to include a steering swivel. The lower control arm is designed with driving thrust rather than braking thrust in mind, and the geometry aimed at giving an anti-squat effect rather than an anti-dive effect (Fig. 8-20).

Space availability in the rear end of the chassis tends to be better than at the front, which gives additional freedom for the engineer to install long locating arms, both longitudinally and transversely. On the other hand, installation space for the spring legs may be lacking because of rear-seat configuration, or the disposition of the fuel tank and trunk layout. Chapman struts are better suited for sports cars without a rear seat than family sedans and are hardly adaptable for station wagons.

Wheel geometry can be selected to minimize camber changes—in fact, it is difficult to obtain large camber changes unless the control arms are made very short and the spring legs sharply inclined.

The plane of the control-arm pivot axis will determine toe-in changes as well as the exact pattern of camber variations. Roll center height is calculated as for the MacPherson-type of front suspension, with an instantaneous center at the intersection of the line extended from the lower control arm and the perpendicular line drawn from the top bearing in the spring leg. Consequently, the engineers have great freedom to adjust roll

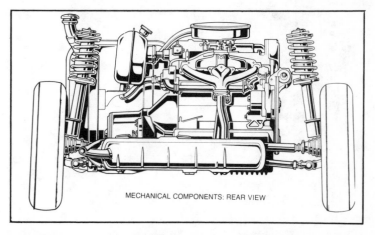

MECHANICAL COMPONENTS: REAR VIEW

Fig. 8-19. MacPherson spring legs are used for the rear suspension of the Fiat X-1/9. Power train and suspension correspond to front end of the Fiat 128, minus steering.

141

center height so as to obtain the desired roll-stiffness distribution and weight-transfer characteristics (Fig. 8-21).

Chapman usually carry coil springs around their upper portions and contain telescopic shock absorbers. The struts also lend themselves well to combination with self-leveling devices.

The deDion Axle

Cars equipped with de Dion axles do not have independent rear suspension, but a system that eliminates the torque-reaction problems of the conventional rear axle. The de Dion axle is a tube or I-beam that connects the two rear-wheel hubs and keeps the wheels in a permanent

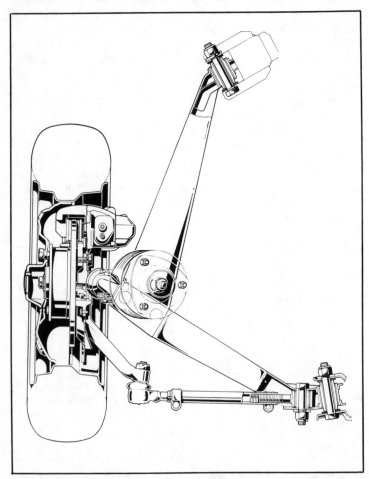

Fig. 8-20. Plan view of the Fiat X-1/9 rear suspension shows the semi-lateral arrangement of the lower control arm and connection to the stabilizer bar.

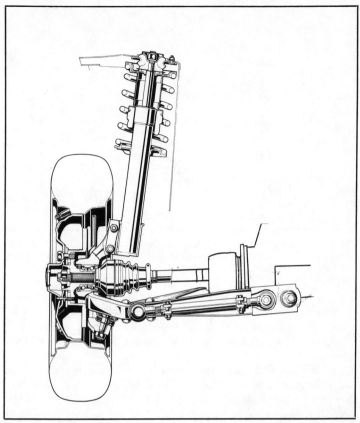

Fig. 8-21. Whether wheels are driven or not makes no difference to the wheel geometry of MacPherson-suspension systems. The Fiat X-1/9 has a high rear roll center, near-constant negative camber, and practically non-variable track.

relationship with each other. The wheels never change from their static camber or toe-in settings (which are usually zero or pretty close to zero) (Fig. 8-22.)

On the other hand, a single-wheel bump will cause both wheels to change camber relative to the roadway. The de Dion axle shares this drawback with other non-independent suspension systems. Its big advantage is having separated the driving torque from the wheel-locating duties into separate components.

The de Dion axle provides a high roll center (at spring anchorage height) and, with regard to unsprung weight, it represents a halfway step between rear axle suspension and independent rear suspension. De Dion suspension systems require double-jointed drive shafts and offer an opportunity for inboard location of the rear-brake units.

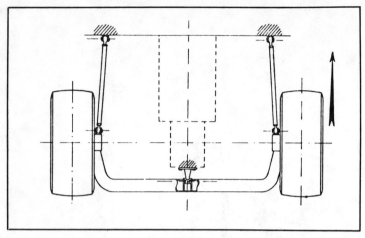

Fig. 8-22. De Dion suspension geometry features constant track and absence of roll camber. Toe-in changes are strictly limited and affect both wheels (as in live-axle suspensions).

Radius arms are needed to take up the driving thrust and generally locate the wheels in the fore-and-aft plane. Lateral location can be achieved by providing the final-drive casing with a vertical slot into which is pegged a bracket pointing straight forward from the center of the de Dion tube.

This is a lightweight solution, often used on racing cars where noise and vibration are not considered as being of any consequence. On production cars, the most elegant solution is a Watt linkage, as used by Alfa Romeo. It assures positive lateral location throughout the full extent of wheel travel and adds little to the unsprung weight.

The geometry of the radius arms determines the roll-steer characteristic of the de Dion axle. Trailing arms minimize the roll-steer effect,

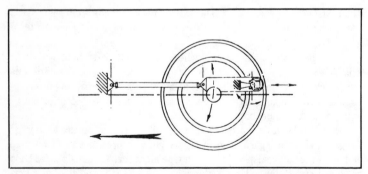

Fig. 8-23. Torque reactions are eliminated in the de Dion-type of rear suspension, and driving thrust is taken up by the radius arms.

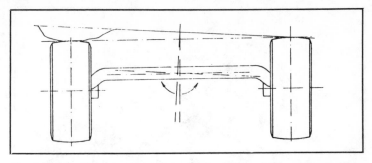

Fig. 8-24. De Dion tube makes wheels non-independent and keeps them always vertical to the road surface, but they are subject to camber changes on single-wheel bumps.

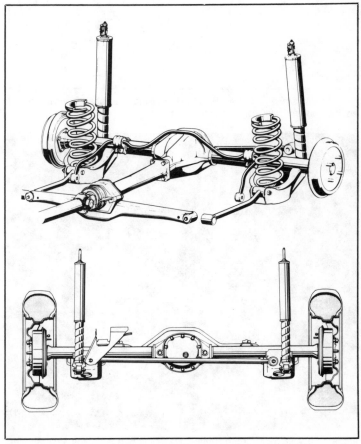

Fig. 8-25. Opel Ascona of 1975 adopted the short torque tube with an interesting application of radius rods at the head of the trailing arms.

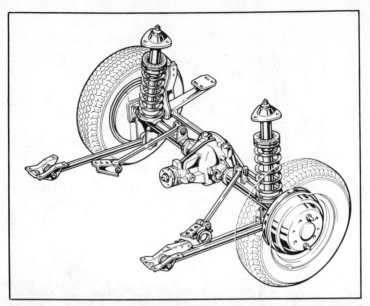

Fig. 8-26. With the Fiat 131 of 1974, the torque tube was abandoned in favor of an elaborate five-link suspension system.

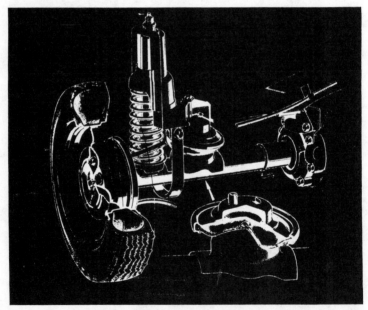

Fig. 8-27. Swing axles of the Renault Dauphine had no radius arms and the driving thrust was taken up by the bearings that housed the universal joints. This version shows the Aerostable compensating air spring (Gregoire patent).

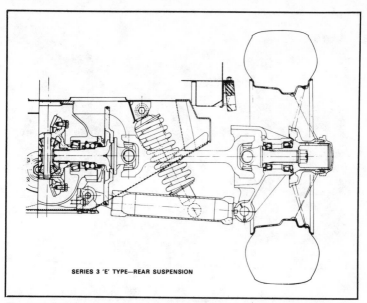

Fig. 8-37. E-Type Jaguar rear suspension has dual coil springs, one ahead of the other, behind the stress-carrying drive shaft. Lower control arms take up the driving thrust.

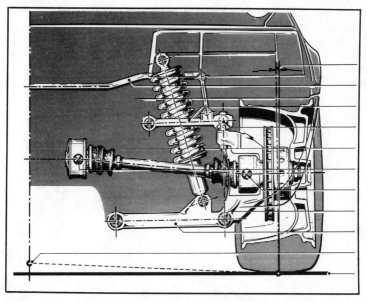

Fig. 8-38. Coil spring in rear suspension of the BMW M-1 runs through the upper control arm. Drive shafts are not stress-carrying members, but simply transmit torque. Roll center is about one inch above ground level.

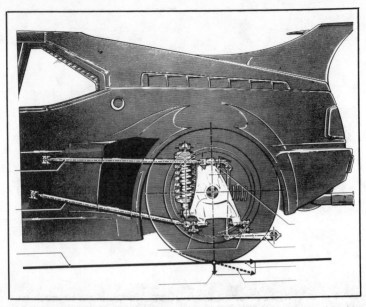

Fig. 8-39. Radius arms in BMW M-1 rear suspension are angled to provide maximum anti-squat effect, aligning their instantenous center to coincide with the line of weight transfer.

low-cost and minimize the sprung weight as far as possible without going to inboard brakes (which would mean adding shafts to turn the discs or drums). Lateral A-arm systems are common, and Chapman struts are also used on front wheel drive cars.

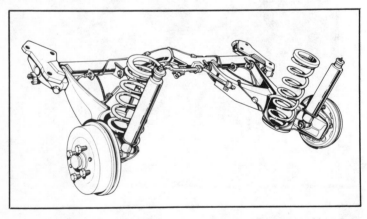

Fig. 8-40. Trailing-arm types of rear suspension started with non-driven rear wheels. This design from the Simca 1307/1308 is fully adaptable to driven wheels.

154

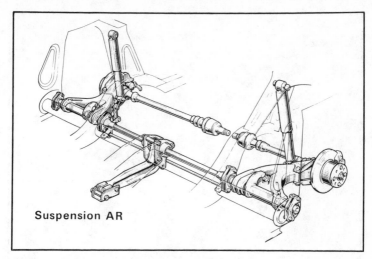

Suspension AR

Fig. 8-41. Driven rear wheels of the Matra-Simca Bagheera are carried on trailing arms linked to transverse torsion bars.

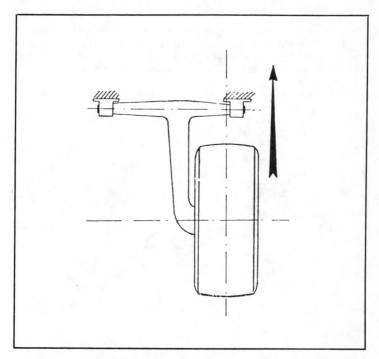

Fig. 8-42. Wheels on trailing arms run like wheels on casters. There are no track variations, and camber angles follow body roll. Roll center is at ground level. There are no toe-in changes during deflection.

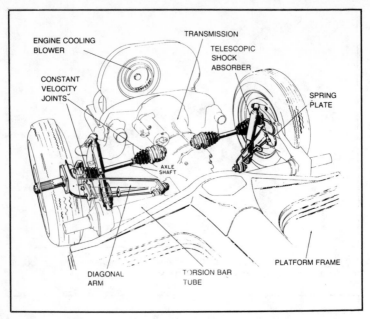

Fig. 8-43. When Volkswagen gave up on swing axles, a system with semi-trailing arms was developed (while Porsche adopted straight trailing arms for its rear suspension on the 901 and 911). Volkswagen remained faithful to torsion bars.

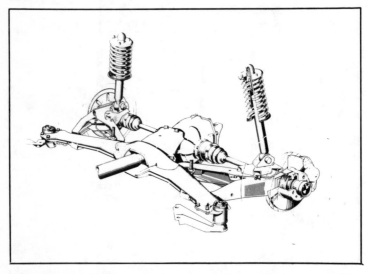

Fig. 8-44. Latest thinking at BMW is apparent from the rear suspension of the 7-series with non-guiding spring legs mounted on semi-trailing arms. Disc brakes are carried at the wheels.

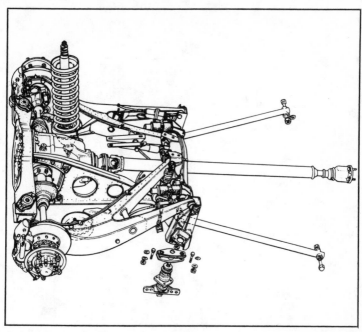

Fig. 8-45. Rolls-Royce went to independent rear suspension for the Silver Shadow and chose semi-trailing arms with a narrow base and narrow pivot angle.

Fig. 8- 46. The semi-trailing arms of the Peugeot 504 are pivoted on a sub-frame whose ends are mounted in resilient rubber blocks bolted to the body shell. This provides exceptionally good insulation from noise and vibration.

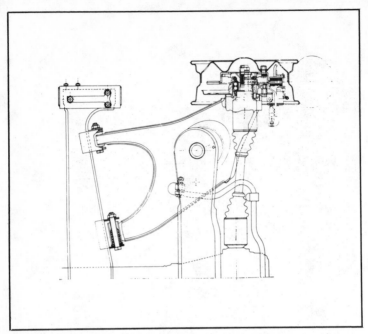

Fig. 8-47. Viewed from above, the Peugeot 504 rear-wheel geometry shows a backwards bend in the drive shaft in the static position. The pivot angle is a mere 11 degrees from transverse (trailing).

Fig. 8-48. Mercedes-Benz semi-trailing arm suspension has a low roll center and uses a hefty stabilizer bar. Semi-trailing arms are carried on the boomerang-shaped sub-frame.

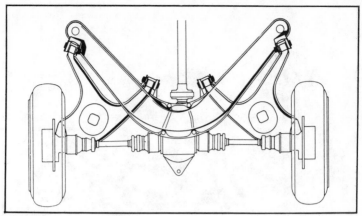

Fig. 8-49. Drive shafts on Mercedes-Benz cars are straight in the neutral position of semi-trailing arms, which are mounted with a 22-1/2 degree pivot angle.

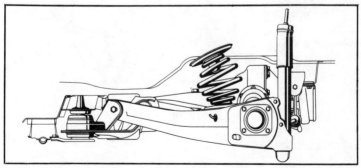

Fig. 8-50. Zero camber at design height is a basic feature of the Mercedes-Benz independent rear suspension. Wheels go into negative camber with slight toe-in jounce.

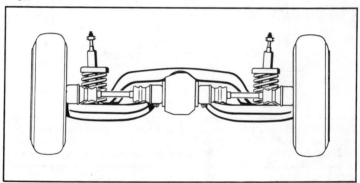

Fig. 8-51. Double-conical coil springs of Opel Senator and Monza are very short and inclined forward to conform with the arc of the semi-trailing arms.

159

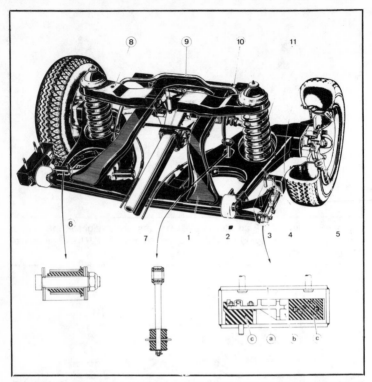

Fig. 8-52. Semi-trailing-arm rear suspension of the Peugeot 604 is developed from that of the 504 with extended use of rubber blocks and bushings. Stabilizer-bar design and linkage are a clever solution to a space problem.

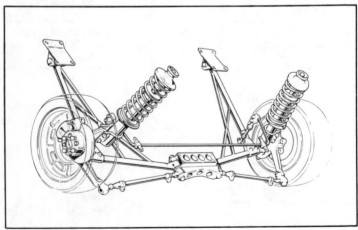

Fig. 8-53. Control arms of a length usually found only on open-wheel racing cars are used for the MacPherson-type rear end of the Fiat-Abarth 131 sedan.

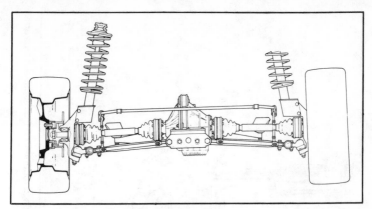

Fig. 8-54. Fiat-Abarth 131 rear suspension gives a fairly high roll center (three inches above gound level) and has a stabilizer bar attached to the lower control arms.

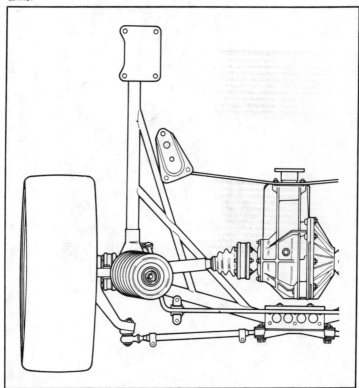

Fig. 8-55. Plan view of the Fiat-Abarth 131 rear wheel and suspension show steering arm and adjustable drag link for easy resetting of toe-in to suit particular conditions.

161

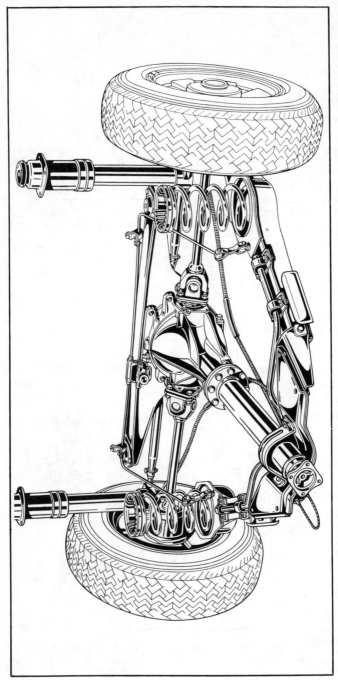

Fig. 8-56. Guiding spring legs without coils served as MacPherson struts on the Fiat 130 where coil springs, mounted on the diagonal radius arms, carried the load. Unsplined drive shafts completed the triangle of lower control arm.

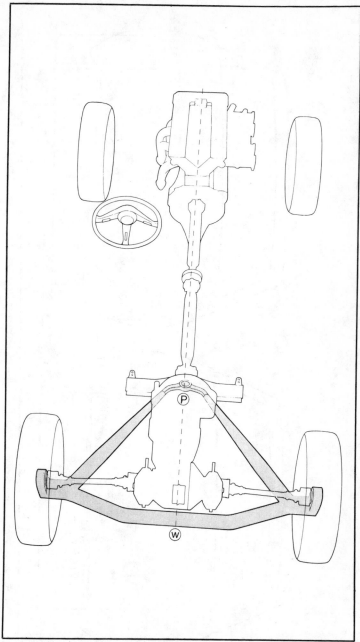

Fig. 8-57. On the Alfetta, the de Dion tube forms a triangle with the two diagonal radius arms. The whole assembly remains perfectly centered by the front Pivot point P and the action of the Watt linkage W behind the tube.

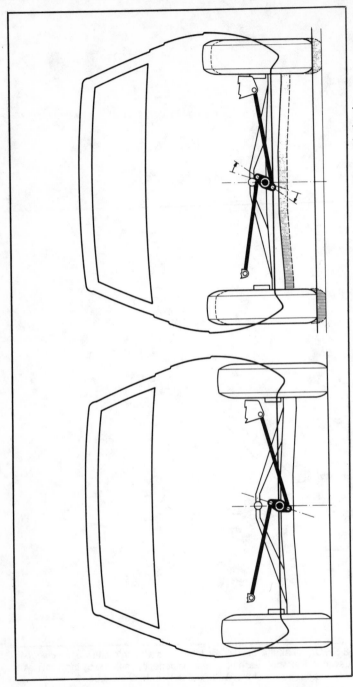

Fig. 8-58. Watt linkage at rest (left) and in jounce (right) assures positive lateral control of the road wheels.

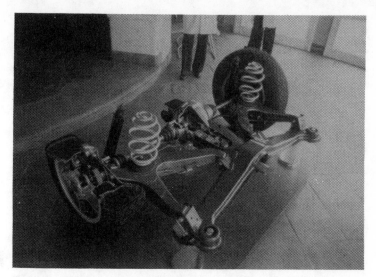

Fig. 8-59. Latest trends are exemplified in the Opel Senator rear suspension with very wide-based semi-trailing arms on a relatively slight pivot axis.

Whenever roll center height seems inadequate (it's at ground level with trailing arms, remember), the engineers can increase roll stiffness by adding stabilizer bars.

Chapter 9
Automatic Level Control

Strange as it may seem at first, soft springs, chosen for their ability to improve ride comfort, can in themselves put a forbidding obstacle in the way of a soft ride. The reason lies in the load variations encountered in a family car.

If rear spring rates are selected to provide a pleasant ride for the driver who is alone in the car, the load of a few passengers and their luggage will use up most of the available jounce travel before the car even starts moving down the road. That means bottoming will occur, even on slight roadway unevenness when the car reaches cruising speed. That's uncomfortable for the occupants and can cause damage to the suspension system. It also puts the car into a nose-up attitude. The front suspension moves into rebound with accompanying changes in wheel alignment that are certain to increase tire wear, slow down the steering response and diminish the handling precision. These attitude changes can also have a damaging effect on the car's aerodynamics. Raising or lowering the front or rear ends will move the stagnation line and materially alter the air flow pattern and the car's lift characteristics. With cars running in a nose-up attitude at night, their headlight beams are pointed not on the roadway, but at the sky—an indisputable safety hazard.

Trucks, where load variations far exceed those of cars and station wagons, have springs designed to handle maximum loads with frank disregard for ride comfort. And that would not be acceptable in a modern car.

Auxiliary Springs

Instead, chassis engineers have been forced to go beyond the simple compromise between spring rates and load variations that take away a little bit of ride comfort in order to keep the car's attitude under better control. The simplest of such systems involve the use of auxiliary springs that go into action after an advanced point in jounce travel only. With leaf springs, such a device consists quite simply of an extra leaf (or more) positioned on top of the main leaf, but of shorter length so that they have no effect on small jounce movements, but abut against a frame member on continued deflection.

Auxiliary coil springs, wound around the shock absorbers and working in compression, are difficult to install so that there is an adequate delay between initial jounce deflection and auxiliary spring action. These *load-levelers* add to the spring rate, with consequent stiffening of the ride under light-load conditions.

Coil springs in tension can be arranged to do the same job while adding less to the unsprung weight. Visualize a pair of horizontal coil springs installed on either side of the wheel hub, one pointing forwards and the other backwards. They undergo stretching on both jounce and rebound travel, and offer progressively stronger resistance to further deflection towards the extremes of wheel travel.

Delco's Auxiliary Air Chamber

In recent years, non-metallic media have come into general use for automatic level control. The most popular systems work only on the rear suspension because that's where load variations are felt most and where the most can be accomplished by relatively low-cost means.

In view of the growing complexity and sophistication of shock absorbers, it was a relatively small step in thought to add the idea of fitting an external air chamber on top of the hydraulic damper. The first and best-known of such systems is the Delco, used as standard equipment on certain Cadillac models since 1965.

The system consists of an air compressor, a reservoir tank, pressure regulator assembly, height control valve, link, special shock absorbers and flexible air lines. The compressor is a two-stage, vacuum-actuated type, requiring no lubrication. Vacuum supply is taken from the engine carburetor base. The compressor's first stage draws air at atmospheric pressure through a one-way check valve. The second stage feeds high-pressure air to the reservoir tank.

On the first-stage compression stroke, the intake valve is closed; the check valve in the second-stage end of the piston is opened. This allows air from the first-stage cylinder to flow out through the hollow piston into the second-stage cylinder for high pressure compression. The second-stage compression stroke closes the check valve in the piston and opens the check valve in the end of the second-stage housing.

Intake and compression strokes are controlled by a sliding distributor valve. The valve is actuated through an arm that is tripped by the piston as it nears the end of each stroke. Each time the arm actuates the distributor valve, a different set of holes is covered in the first-stage housing. The distributor valve controls the flow of intake manifold vacuum and air under atmospheric pressure—alternately on opposite sides of the compressor diaphragm.

As the compressor runs, the reservoir air pressure gradually builds up. This gives rise to a back pressure on the second-stage piston until it equals the push or pressure against the diaphragm. At this point, the system is in a balanced condition and the unit stops operating. When action in the system reduces reservoir pressure, the compressor resumes its function and refills the reservoir. Pressure balance will depend on the prevailing manifold vacuum and atmospheric pressure. Both are affected by changes in altitude (above sea level). The pressure regulator valve is preset and limits the reservoir outlet pressure to avoid damage to the height control valve and shock absorbers.

The height control valve is mounted on the frame. It senses rear car height through a link attached to the right-rear upper control arm. When load is added, the overtravel level is forced up. This brings an internal lever to open the intake valve. When this valve is open, high-pressure air is admitted to the shock absorber air chambers. As the car is raised, the intake valve shuts. When load is removed, the overtravel lever is forced down, causing the internal arm to open the exhaust valve. When the car is lowered to level position, the exhaust valve shuts. To prevent air transfer under normal road movements, a four- to 15-second delay mechanism is built into the height control valve. The overtravel lever, which pivots around the control valve shaft, rides off the flat side of the control shaft and does not have time to react to the high-frequency changes of ride motions.

For future cars, where manifold vacuum will not be available for running the compressor (due to new air-pollution control systems), Delco plans to use pneumatic levelers with an electrically driven compressor and an electronic control system.

Motorcraft (Ford), Monroe, Goodyear, Armstrong, Hoesch, Koni and Boge have developed and are manufacturing rear-end, self-leveling systems. Since 1965, Rolls-Royce and Bentley have had their own, based on the Delco principles. Mercedes-Benz has used Boge hydro-pneumatic levelers as an option for its smaller series since 1967, and a new Boge all-hydraulic device is standard of the T-series wagons. Rover also chose Boge equipment for the 2500 and 3500 sedans.

Gas/Hydraulic Suspension Systems

From two-wheel (rear) level control, it is a logical step to extend the system to comprise all four wheels. And the more you consider this idea, the more inevitable the conclusion becomes that the self-leveling system should be used not as an adjunct to the metal springs, but as their replacement.

This line of thought leads to the four-wheel gas/hydraulic suspension system, in which a compressed gas acts as the elastic element and a hydraulic fluid carries the load and provides the damping. Today, only two companies produce cars with gas/hydraulic self-leveling suspension systems—Citroen and Mercedes-Benz (Fig. 9-1).

Hydragas System on Austin

In addition, the Austin Allegro has a gas/hydraulic suspension system, but it does not provide true level control, though it is designed to minimize attitude changes due to load variations (Fig. 9-2). It is called *Hydragas* and is aimed at giving a soft and pitch-free ride with constant spring rates. It also provides high enough roll stiffness to obviate the need for stabilizer bars.

Hydragas suspension is essentially simple, comprising an integral spring and damper unit at each wheel. These units use a sealed-for-life inert gas as an elastic medium. The weight of the car is carried by a water-based fluid under pressure and the units are interconnected front-to-rear, the whole system being hermetically sealed (Fig. 9-3).

The Hydragas unit is a nitrogen-filled, spherical chamber welded onto the top of a dispenser chamber. Between these two chambers are carefully shaped holes covered by rubber compression blocks to control the flow of fluid between the upper and lower chamber. This valve is a two-way device and provides the required suspension damping.

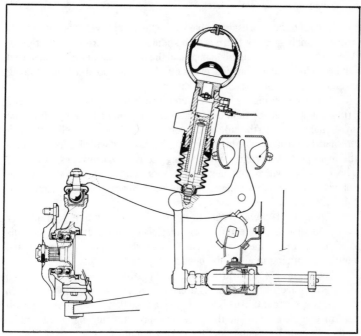

Fig. 9-1. Hydropneumatic spring unit of Citroen GS works with short travel from its base near inboard end of upper control arm. Stabilizer bar is connected near the same spot.

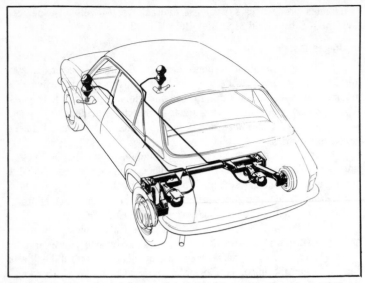

Fig. 9-2. Hydragas installation on the Austin Allegro has direct hydraulic inter-connection without central control.

The rubber compression valve within the unit provides the necessary bounce damping. Since there are no sliding seals, there is no friction or wear. This allows car-life damping, and the large quantity of water-based fluid which circulates avoids excessive heat buildup, thus maintaining constancy of damping arduous conditions.

At low velocities the flow in the rebound is confined to the main bleed orifice. In jounce, the fluid flows through the bump bleeds by lifting the small rubber flaps on the side on the compression block, as well as through the main bleed orifice. At higher velocities, the fluid will also flow through the main ports by progressively lifting the rubber compression blocks. The housings which surround the blocks control the amount of pre-compression that is applied to the rubber and, hence, the flow characteristics.

The pitch damping occurs in the interconnection pipe system. For each unit pipe is connected to the displacer (lower chamber) and this configuration is termed underport. This allows completely separate and optimum damping in pitch and bounce.

Hydragas operates with a leverage of between four and five to one to reduce the size of the units. The cross-tube rear suspension has been designed to contain these high-levered loads so that only forces of wheel-load magnitude are fed into the body structure. It is rubber-mounted at its ends to a compliant subframe of extreme simplicity.

The extent of softening the wheel rate in the pitch mode is constrained by consideration of trim change with load. Limitation of trim changes is mainly achieved by the torsional rate of the dual concentric bonded rubber bushes in the rear trailing arms. On the Allegro, these rubber bushes are free when the wheel is in jounce position. In the unladen position, they provide a false load at each wheel of 81 pounds, pulling the body down. This false load reduces the percentage increase of load at the rear when the car is laden with passengers and luggage and has the effect of reducing the attitude change.

At the front, the units are mounted vertically with the piston acting on the upper lateral A-arm. The upper control arm is pivoted on two 0.75-inch bore dual concentric bushings which take the unit load, provide the required parasitic rate and allow some degree of horizontal compliance.

Oil Carries Load in Citroen/M-B Systems

Citroen uses gas/hydraulic suspension systems on two series in its current model lineup—the GS and the CX (Fig. 9-4). The systems differ in dimensions and detail design, but are based on the same principles and will be dealt with as one design. The design provides fully automatic level control, but depends on conventional stabilizer bars for roll stiffness.

Ride frequency variations are carefully controlled and are among the lowest obtained in production cars. The CX has a range from 35.7 cycles per minute unladen to 38.4 with a full load, at the front and 31.2 (unladen) to 45.6 (full load) at the rear.

Spring rate variations are made to accommodate important changes in load. Since these changes occur mainly at the rear end, the rate varies from

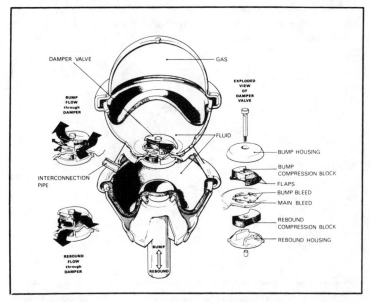

Fig. 9-3. Each Hydragas unit contains a rubber spring, damper units, fluid-level control chamber, and gas-filled variable-volume spring element.

61.9 (unladen) to 82.9 (full load) pounds per inch for the front suspension on the CX. But at the rear, spring rates vary between 190 (unladen) and 86.2 (full load) pounds per inch. Combining such characteristics with unusually generous wheel travel enables the Citroen engineers to claim ten times greater elasticity in the suspension system of their cars than in the average car using metallic springs.

Both GS and CX have four-wheel independent suspension with lateral A-arms at front and trailing arms in the rear. The front spring units stand, inclined inwards and backwards, on top of the upper control arm, while the rear spring units are placed horizontally inside the frame side members.

Each spring unit consists of a cylinder and a sphere. The piston inside the cylinder is connected to a suspension arm-directly at the front and via a bell-crank at the rear. The piston works against a volume of hydraulic fluid (a mineral oil) that fills the upper part of the cylinder and part of the spherical chamber. The other end of the sphere is filled with gas. The gas and the fluid are separated by a diaphragm.

In jounce, the piston displaces fluid from the cylinder into the sphere, which compresses the gas. The gas works as a pneumatic spring (think of it as an air mattress). Whenever the road wheel goes into jounce , the volume of gas is reduced.

But that does not make the car self-leveling. Its ride height is kept constant by variations in the volume of imcompressible fluid between the piston and the diaphragm, and these variations are made by a pressure regulator mounted on the spring unit cylinder, connected via hydraulic lines to a high-pressure pump, a storage tank and a safety valve with front and rear corrector valves.

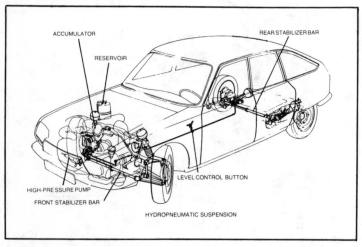

Fig. 9-4. Citroen GS self-leveling suspension with air-and-oil spring units at all wheels is controlled by a central hydraulic system. Front A-arms and rear trailing arms give roll axis at ground level.

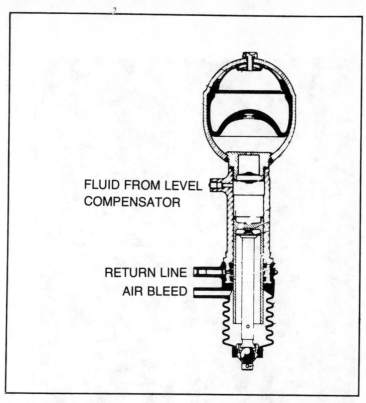

FLUID FROM LEVEL COMPENSATOR

RETURN LINE
AIR BLEED

Fig. 9-5. Lower part of Citroen spring unit contains a shock absorber; upper part has height corrector. Sphere on top contains air above the diaphragm, hydraulic fluid below.

The Citroen suspension works in a reversible mode. As the load is increased, the chassis dips, causing an angular movement of the control arms. This motion is linked via the stabilizer bar to a control rod which actuates a sliding valve inside the pressure regulator at the same time the piston begins to displace the fluid inside the spring unit for that wheel. Opening the sliding valve establishes a circuit between the hydraulic fluid under pressure and the cylinder. Due to the influx of additional fluid, the chassis is raised to its normal height. The return movement of the control arm returns the valve to its neutral position and shuts off the fluid flow. Conversely, when load is reduced, the body rises because the gas is relieved or some amount of pressure. At the same time, the mechanical linkage to the sliding valve turns to its return position and allows fluid to flow back to the reservoir. The reduction in fluid volume then restores the cars level to neutral.

Fluid flow between the cylinder and the sphere must pass through a hydraulic shock absorber (Fig. 9-5), which works throughout all cycles

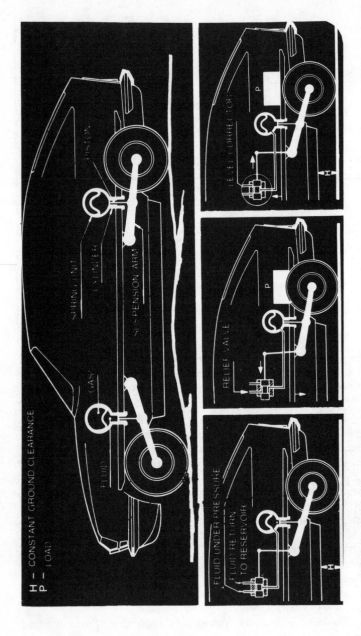

H — CONSTANT GROUND CLEARANCE
P — LOAD

SPRING NUT
PISTON
CYLINDER
GAS
FLUID
SUSPENSION ARM

RELIEF VALVE
P

LEVEL CORRECTOR
P

FLUID UNDER PRESSURE
FLUID RETURN TO RESERVOIR

Fig. 9-6. Schematic presentation of how the Citroen suspension assures constant level control—front/rear interconnection is purely hydraulic.

174

with positive internal pressure so as to eliminate cavitation in the fluid. In addition to acting as a normal damper on wheel travel, the shock absorber filters out vibrations from the wheels and, to some extent, holds back the wheel deflections from reaching full amplitude.

The Citroen system includes further refinements such as a manual height-control lever which enables the driver to select one of three levels of ride height (high for maximum ground clearance on rough roads, low for highway driving). The lever acts directly on the corrector circuit and does not affect spring rates or ride frequencies (Fig. 9-6).

In addition, the hydraulic suspension lines are part of a central high-pressure hydraulic system which also supplies the power assist for the brakes and the steering. The height correctors also sense dynamic changes in load during front-and-rear weight transfer and, thereby, provide the means for effective brake-force distribution according to wheel loading.

After many years' experience with air springs (see the evolutionary chapters), Mercedes-Benz developed its own hydro-pneumatic suspension for the 1975-model 450 SEL 6.9. The hydro-pneumatic Mercedes-Benz system follows the Citroen principles, with air as the elastic element and oil to carry the load and do the damping. Mercedes-Benz has some refinements of its own, including provision for isolating the spring units to prevent sagging when the car is parked for long periods.

An upright telescopic strut incorporating a hydraulic shock absorber replaces the coil spring at each wheel and is connected to a central hydraulic pressure tank. An engine-driven pump pressurizes the reservoir, which feeds individual circuits for each wheel.

Chapter 10
Where It All Began

Of course, it all began with the wheel. The actual origins and invention of the wheel are not known but have been traced to the Orient. Tribes of Asiatic nomads are believed to have been the first to use the wheel for transport vehicles, which could help explain the widespread use of wheels at a very early date.

Ancient Indian manuscripts dating back to about 1,700 BC mention wheels. And in Holy Writ are references to Egyptian chariots—two-wheeled vehicles—in the period 1715-1705 BC.

Competent historians agree that the wheel probably originated from rollers, made from tree-trunks and placed under heavy loads to enable a crew of men to move them. Once the principle had been established, it was natural that progress would take the course of first cutting up the rollers and then increasing wheel diameter, leading to composite wheel structures made by fastening planks onto the surface of two or more tree-trunk sections, and eventually to multi-piece wheels with a hub, a rim and a number of spokes.

Wheeled vehicles had been around for centuries before suspension systems came into being. On the Egyptian, Persian, Assyrian and Grecian chariots, the axles were mounted rigidly to the frames; when two-axle vehicles were introduced, carriage-makers adopted the same primitive type of mounting.

The First Suspensions

Springs were introduced on medieval carriages not because of road-holding considerations, but solely because the passengers began to demand a certain ride comfort. The *carrosse* of the 16th century had its axles mounted rigidly on a more or less flexible frame, while the passengers sat in a compartment that was suspended, first by iron chains and later by leather straps, from four posts rising from the front and rear ends of the frame.

It was not until 1750 that metal springs became available to the carriage builders. They did not, however, interpose springs between the axles and the frame; they merely replaced the mounting posts for the passenger compartment with long leaf springs, retaining also the leather straps.

In 1805, and English coachbuilder named Elliott was the first to install springs as axle suspensions—with full-elliptic leaf springs placed longitudinally at all four wheels. The front axle was pivoted on the frame at its center so that the whole fore-carriage turned on curves.

The steam carriage boom in England between 1820 and 1840 led inventors to concentrate of the drive train, however, and not on the suspension system. William Charge's 1832 patent for a steam vehicle with four sprung wheels and chain drive received little attention until many years later.

The true originators of the automobile, Daimler and Benz, were curiously unconcerned with suspension systems. Gottlieb Daimler took no interest in chassis design, as his purpose was to build and sell engines for installation in existing carriages, boats and stationary power equipment. He did not envisage a need for anyone to drive cars at speeds in excess of 12 mph. Carl Benz was also mainly an engine man and avoided steering and suspension problems by making his first car a three-wheeler.

Maybach and Bolee Designs

It was Daimler's assistant, the gifted Wilhelm Maybach, who first saw the automobile chassis as something completely different from a horse-drawn carriage, seeking inspiration instead from bicycle engineering; but his first effort in this field displayed remarkable ignorance of the problems involved with cars. Maybach very intelligently designed a lightweight steel-tube frame for the 1889 Stahlradwagen, and the front wheels were mounted on forks, motorcycle-fashion (Fig. 10-1).

The wheels were large-diameter with steel wire spokes. The front forks and rigid rear axle were solidly attached to the frame. Perhaps as an afterthought, Maybach decided to use springs somehow; but, unfortunately, they did not go into the chassis-the four coil springs went under the four corners of the bench seat!

With all due respect to Maybach for the lightness of the Stahlradwagen, it must be concluded that more advanced suspension principles had been applied 16 years earlier on a heavy vehicle made by Amedee Bollee, Sr. in the mechanical workshops he had started as an annex to the family's iron foundries at Le Mans. His L'Obeissante steam carriage of 1873 was a 12-seater weighing about four tons, built on a true perimeter frame which carried mounting brackets for the suspension springs of all four wheels. At

Fig. 10-1. Maybach's Stahlradwagen was a product of bicycle engineering, rather than carriage-building, and the chassis has no suspension system—only the seat rested on coil springs.

the front end, the wheels were carried on spindles whose ends were fork-guided and supported on full-elliptic leaf springs. Here, in fact, we have the world's first independent front suspension system! At the rear end was a driving axle, suspended by semi-elliptic leaf springs carried railway-fashion outboard of the hubs.

As Bolee refined his theories, he began the construction of his second steam vehicle, La Mancelle, which was completed in 1878 (Fig. 10-2). This was a much smaller machine, resembling a horse-drawn victoria and weighing less than two tons.

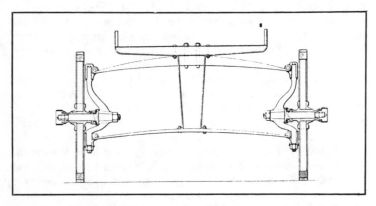

Fig. 10-2. Upper and lower transverse leaf springs linked to tall steering swivels gave the 1878 LaMancelle a modern transverse-arm type of independent front suspension.

La Mancelle featured a brilliant system of independent front suspension with equal-length A-arms made up of upper and lower transverse leaf springs. The leaves were tapered towards the ends, having very wide center sections, so that they were able to assure the fore-and-aft location of the wheels as well as guiding the wheels during spring deflection.

Amedee Bollee, Sr. built a number of steam vehicles after that, but opted for I-beam front axles. However, his son, Leon Bollee, revived the front suspension design of La Mancelle for a gasoline-driven car he built in 1898.

Varied Springing with Front Axles

When car-building became an industry, most manufacturers avoided experimentation and settled for simplicity. What has become recognized as the world's first series-production model, the 1893 Benz Velo (Fig. 10-3), had an I-beam front axle with full-elliptic springs-a layout that was widely copied.

As early as 1891, Peugeot had built a car combining a transverse leaf spring with an I-beam front axle, but later adopted semi-elliptics (Fig. 10-4). The 1901 Mercedes, designed by Wilhelm Maybach, also used semi-elliptics which then became the industry norm. But the transverse leaf was not forgotten. Ford adopted it for the Model T in 1908, and it was retained on the Model A, Model B and all V-8's through 1948.

While European engineers shied away from using coil springs, an American pioneer, Alanson P. Brush, used coil springs on the front axle of the Brush Runabout of 1907. It is noteworthy that, in contrast with what later became normal practice, Brush used the coil springs in tension; the body-and-frame weight was carried by a bracket that extended to the coil spring base. This base had a vertical rod, concentric with the spring and

Fig. 10-3. Front axle on 1893 Benz was sprung by means of a pair of full-elliptical leaf springs tied to the frame top and bottom for stability.

Fig. 10-4. Peugeot was the first to use a transverse leaf spring linked to the front axle, starting in 1891. This is an 1894 version.

with a lid on top, securely fastened to the uppermost coil. The lower end of the rod pivoted in a socket on the front axle. A bump would move the rod upwards, and the lid would stretch the spring. The axle was located by twin radius arms on each side which also worked as levers for friction-type shock absorbers.

Front Brakes Brought New Problems

The average car, European and American, lived happily with the front axle until the arrival of front brakes, which were adopted industry-wide between 1920 and 1930. This caused problems, especially for axles carried on semi-elliptic springs.

First, this front-axle problem was due to the braking effect itself; throwing horizontal loads into springs meant to work only in the vertical plane. Second, it was due to the weight of the brake drums, carried as unsprung weight in the front wheels and materially increasing the moment of inertia of the entire front-axle assembly. In practice, the drums lowered the natural frequency of axle oscillation around its geometrical center. The driver felt such oscillations as shimmy—the front wheels steering alternately left and right in a repeated cycle that could not be checked by muscle power.

At the same time, engines were becoming more powerful and the cars could reach higher speeds. Also in the 1920's, low-pressure balloon tires were standardized by all leading auto makers. They tended to reduce wheel hop frequency, which in some cases, became low enough to set up axle oscillations at its natural frequency. Though the tires gave far better ride comfort, they added to the hazards of driving, even on moderately bad roads.

To combat these problems, the manufacturers stiffened the frames and reinforced the cross members, leading the way to the heavy, cross-braced frames of the 1930's. Some engineers fitted a kick shackle on the front spring adjacent to the steering gear so as to limit the shimmy forces fed back to the steering wheel. In France, both Farman and Cottin-Desgouttes produced expensive solutions with a separate steering gear for each front wheel, thus eliminating the trackrod that played a part in the transfer of shimmy reactions.

Engineering Struggle Toward IFS

But these were only half-measures, and not truly effective. All over the world, automotive engineers were looking for a solution. More and more chassis experts began to study independent front suspension. If the mechanical connection between the two wheels was broken, they reasoned, the shimmy problem would disappear. It was an old idea, but the engineers that had used such designs before the turn of the century, or in the years prior to World War I, had done so for reasons unconnected with the problems of the heavy cars of the late 1920's. Now they were greeted with new respect.

Most of the engineers and researchers were working in darkness, unaware of the effects their design changes would have on the car's dynamic behavior. Even the most imaginative and daring chassis designers were ignorant of certain aspects of independent front suspension as compared with I-beam axles. Some surprising examples will be mentioned later. To avoid confusion, we will look at each type of independent front suspension separately in the hope that the chronology will still be apparent at the end.

Despite the interdependence of front and rear suspension systems of the same car, and the need for matching them to each other, I have chosen to treat the rear end quite separately in the following chapters, while not neglecting to point out certain front ear combinations whenever relevant to giving the reader a full understanding of the car under discussion.

Chapter 11

Development of
Independent Front Suspension

As we have seen in the previous chapter, one of the most advanced types of independent front suspension originated at a remarkably early date. As a corollary, it should be added that some of the more primitive types were not created until quite late.

Even more important is the observation that many early independent front and rear suspension designs were grossly inadequate for handling the locating duties and failed to provide sufficient strength or accurate guidance of the wheels. That's one reason why so many production cars stuck to the rigid front axle so long. Another is that cars with stiff springs on I-beam axles, such as Bentley, Bugatti, Invicta, Delahaye, Maserati, Delage and countless others, had very high handling precision and safe, predictable road behavior.

These and other stiffly sprung sports cars handled very well indeed, but it was more due to their high roll frequency, with resultant quick-steering response and low -roll amplitudes, than to a superior principle. But stiff springs bring on new problems. For instance, they reduce tire adhesion by resisting deflection and, thus, often cause the tires to lose contact with the road surface entirely. In consequence, reasonably low spring rates are a necessity for proper handling characteristics as well as ride comfort.

Shimmy and other control problems that resulted from the combination of soft springs with I-beam front axles, heavy brake drums and balloon tires served more and more to underline the necessity of going to independent front suspension. And a variety of such systems were known at an early point in history.

Sliding Pillar System

What is meant by a sliding pillar is a type of suspension in which the front-wheel spindle carries an extension that fits inside a vertical cylinder.

Each front wheel is able to travel up and down without change in track, wheelbase or camber. The design keeps unsprung weight extremely low, for the cylinder is part of the chassis structure. The cylinder serves as king-pin and provides freedom for the spindle to swing back and forth as directed by the steering arm. The cylinder can be made to contain a coil spring and concentric shock absorber. Alternatively, it can act as a sleeve for a rod that works against the end of a transverse leaf spring. The roll center, of course, lies at ground level.

The inventor of the sliding pillar suspension was a French engineer, F. Cornilleau, who was engaged by Joseph Guedon, Owner of the Decauville works in Bordeaux, to design a light car for series production in 1896.

The Decauville used a transverse leaf spring whose ends were linked to the sliding pillars (Fig. 11-1). Cornilleau patented his invention, and in 1898 Decauville sold licenses for the complete car to two companies determined to form part of the auto industry: Wartburg in Germany and Orio & Marchand in Italy. Both manufactured exact copies, including the independent front suspension.

The first to combine the sliding pillar with coil springs was a remarkable American engineer, famous for his all-wheel-drive and tracked military vehicles, John W. Christie. On his front wheel drive car of 1904, the wheel hubs were attached to two fat towers which contained strong coil springs (but did not permit a large amount of wheel travel).

An English car, built in 1898 by R. J. Stephens, used sliding pillar suspension with a transverse leaf spring, but never got into production. A variant of the Decauville design showed up on the 1906 Sizaire-Naudin, the work of Maurice Sizaire, an artistic draftsman who had worked for an

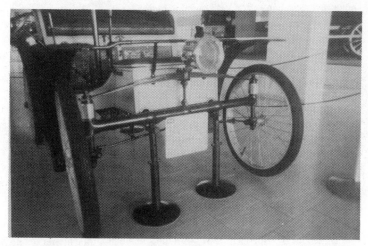

Fig. 11-1. 1898 Decauville voiturette featured independent front suspension with sliding pillars and transverse leaf spring.

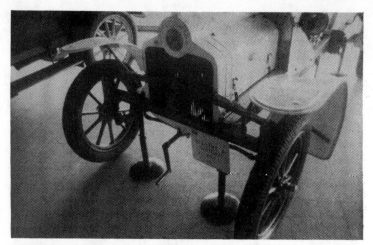

Fig. 11-2. Sizaire-Naudin from 1907 had a transverse leaf spring linked to the upper ends of sliding pillars that held the king-pins.

architect before coming to de Dion-Bouton. Together with Naudin, a machinist who also worked at de Dion-Bouton, he created a new, low, lightweight sports car (Fig. 11-2) that was in production up to World War I.

Up in Gloucestershire, Henry Morgan had no doubt seen the Sizaire-Naudin, though he probably had no knowledge of Christie's pioneering work. Morgan began building three-wheeled sports cars in 1909 and modified the Sizaire-Naudin design to use coil springs instead of a transverse leaf. The springs were narrow coils enclosed in the pillars and worked without benefit of damping. Later, Morgan added friction-type shock absorbers, and after 1945, went to hydraulic ones. Morgan cars still use the sliding pillar suspension.

The Morgan was sold in France in the 1920's as the Darmont. Louis Bechereau built sports cars using a very similar suspension system from 1922 to 1925. But the Clement-Rochelle, built from 1927 to 1932, reverted to the Sizaire-Naudin layout, using a transverse leaf spring.

The fertile brain of Vincenzo Lancia created a new type of sliding pillar suspension, free of outside influence of any kind. During the war, his imagination had been playing with many completely new ideas which were united in the 1923 Lancia Lambda (Fig. 11-3). Most of them, such as the narrow-angle V-four with staggered cylinders, the unit-construction body, and the sliding pillar front suspension were patented.

Lancia included hydraulic shock absorbers in the sliding pillars, along with coil springs. The pillars were oil-filled and sealed. The Lancia design evolved through the years and was used on the 8-cylinder Dilambda in 1928 (Fig. 11-4 and 11-5), Artena and Astura in the early Thirties, and redesigned for the Aprilia in 1937 (with the spindle mounted higher on the pillar, lowering the whole assembly). It was last used on the Aurelia of 1950 to 59, with very sophisticated hydraulics for damping.

184

Fig. 11-3. In 1923 the Lancia Lambda incorporated coil springs and hydraulic shock absorbers in the sliding-pillar front suspension.

Fig. 11-4. Lancia Dilambda from 1928 looked, at first glance, as if it had no suspension at all. Vertical frame members are, in fact, sliding pillars.

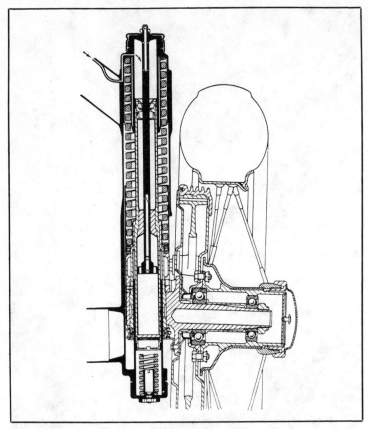

Fig. 11-5. Lancia Lambda sliding-pillar arrangement needed no maintenance, due to oil-filled cylinders, and proved extremely reliable in service.

Only one American make has used a front-suspension design in the sliding pillar family: Nash. It was an unusual design by Meade F. Moore and appeared on the 1940 to 42 model Nash Series 40. Its main component was an inclined spring leg (actually, an enlarged king-pin whose upper portion carried a coil spring. Shock absorbers were not concentric with the spring leg, but mounted separately inboard of the coil springs.

The lower spring-leg mount was an extension of the front cross-member, and the upper mounting was a bracket extending from the body structure. For its post-war models, Nash adopted lateral control arms and coil springs.

Trailing Links System

The technical aspects of trailing links have been discussed in an earlier chapter. Historically they are fairly recent, as they appeared first in a patent issued in 1931 to Dr. Ferdinand Porsche, covering his transverse torsion-bar installation. For this application, the professor preferred laminated torsion bars, which were featured on the Volkswagen Beetle from 1949 to date .

Dr. Porsche used trailing link and torsion bar suspension on a Volkswagen design as early as 1932, but the first production car to make use of the invention was the Alsatian-built Mathis Emysix of 1933, made under Porsche license. Another version of the same design appeared on the Auto Union racing cars from 1934 to 1939, also designed by Dr. Porsche (Fig. 11-6).

Porsche knew that with axle suspensions, the roll center was located at spring anchorage height, but he had no idea that trailing arms have their roll center at ground level. He expected the Auto Union racing car, which had a center of gravity in close proximity to spring anchorage height, to have no roll at all. The fact that body roll did occur made him revise his ideas; but he valued the good points of the trailing links more than he feared their drawbacks, and after the Volkswagen, he never hesitated to use the same Porsche car in 1948. But the trailing links were thrown out when a different engineering team at Porsche brought out the 911 to replace the 356 in 1967. This was an interesting reversal of basic philosophy, for the 356 had a roll axis with a sharp nose-down angle (very high rear-roll center, because of the swing axles). On the 911, the front roll center is about 2-1/2 inches above ground level, while the trailing-arm rear suspension gives a roll center at ground level. Thus, the car rolls about an axis that's down at the tail and pointing upwards at the front. For a car with a high rear bias in weight distribution, the newer arrangement was a far better solution.

In 1935, Alfa Romeo brought out its first car with trailing links, the 6C 2300 B, in combination with vertical coil springs enclosed in a cylinder

Fig. 11-6. Trailing links, early version, are seen on 1936 C-Type Auto Union, with transverse torsion bars and large-diameter friction-type shock absorbers.

which also contained hydraulic shock absorbers. The rear joint of the upper trailing link also had a connection to a lateral arm with its other end attached to a pivot point on the chassis. At its mid-point, a vertical rod with a piston at the lower end ran through the spring cylinder, compressing the spring (in jounce) from the bottom against the fixed abutment at the top.

This system was later used for the 8C 2900 sports car and the 12 C racing car and was retained on the first postwar car, the 6C 2500. But that does not end Alfa Romeo's involvement with trailing links. A new racing car, Tipo 158, known as the Alfetta, appeared in 1938. It had a different design, with less mechanical complication, discarding the enclosed coils in favor of a transverse leaf spring. This design was in use on its successors through 1951.

The ERA E-type, designed by Peter Berthon just before World War II, had a front suspension that was a reasonably frank copy of the Auto Union setup; but only one or two examples of the E-type were built.

In 1946, Donald Healey, who had been tenchical director at Triumph before the war, went into business for himself with a Riley-powered sports car that had a trailing link and coil-spring system of front suspension. Here the power link was a robust radius arm with a spring base near its pivot point. This was used on all Healey cars, including the Silverstone roadster, until 1950 when the coil springs were replaced by torsion bars. Healey dropped the trailing links in favor of lateral A-arms and coil springs, with the advent of the Austin-Healey in 1952.

Claude Hill, chief engineer of Aston Martin, chose a similar design for his 1940 prototype, named Atom, which evolved into the DB-2 of 1946. Here it was the upper link that formed the spring base, with an extension from its rear joint so that the springs were centered on the front wheel axis. On the DB-3 (Fig. 11-7), developed with Prof. Robert Eberan von Eberhorst as a consultant, the coils were replaced by transverse torsion bars and the upper links triangulated for greater resistance to side loads. Eberan von Eberhorst had worked at Auto Union before the war and preferred the Porsche design. But when Harold Beach and John Wyer designed and developed the DB-4, they discarded trailing links altogether and went to lateral A-arms and coil springs.

Peter Berthon of ERA fame was the chief engineer for the V-16 BRM, which first appeared in 1950, and he again opted for a trailing-link type of front suspension. On this car, trailing links were combined with oleo-pneumatic struts. This design was withdrawn in 1953, and when BMW re turned to racing, it was with a four-cylinder car that used lateral A-arms and coil springs in the front suspension.

Dubonnet System

Totally outside all families of independent front suspension, the Dubonnet system contains one element that, in part, defines its geometry: a leading arm. Each front-wheel spindle had its own leading arm—but only

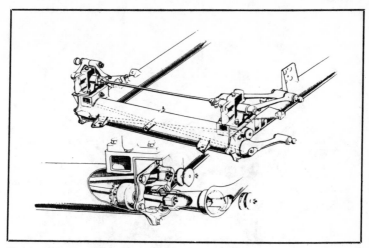

Fig. 11-7. Trailing links, late version, as used on the Aston Martin DB 3, with adjustable torsion bars and triangulated upper links.

one. It's in no way an equal-but-opposite principle to that of the Porsche-type trailing links.

With Dubonnet suspension, the rear end of the leading arm was splined to a pivot shaft mounted across the bottom end of a spring cylinder that was oil-filled and contained a coil spring and hydraulic shock absorber. This whole cylinder was carried on the king-pin and swiveled with the wheel in an exact reproduction of its steering angles. The king-pin, to make matters clear, was removed from the wheel and mounted on a chassis cross-member or simulated axle beam forming an integral part of the frame.

The pivot shaft at the bottom of the cylinder carried a rocker arm that compressed the coil spring in jounce, stretched it slightly in eebound, and also actuated the piston rod of the shock absorber. All wheel travel was absorbed inside the cylinder which, as mentioned above, participated in steering movements—not as part of the linkage, but in order to separate the steering linkage from the effects of spring deflection.

It was an ingenious system which gave a ground-level roll center and eliminated camber changes on single-wheel bumps. Body roll angles, however, were reproduced in the wheels. Depending on the length of the leading arm, it permitted generous wheel travel and low spring rates. Its unsprung weight was extremely low.

It had one geometrical drawback in that it was not free of gyroscopic wheel travel so that the tire footprint center would be displaced relative to the king-pin center line. This also altered the caster action, which was a point of minor importance.

The main weakness with the Dubonnet suspension system lay in its construction and the strength of its moving parts. Adequate bearing areas would be multipled by the length of the leading arm.

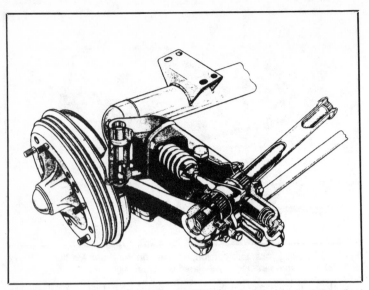

Fig. 11-8. Dubonnet suspension for the 1948 Vauxhall Wyvern and Velox had short transverse torsion bars, plus an auxiliary coil spring inside the cylinder.

In its original form, the Dubonnet suspension system was far a more complex that the description above, which relates to the American production version. In the patent drawings from 1931, we find three sealed cylinders, each with separate coil springs for jounce and rebound. Dubonnet built prototypes with such assemblies from about 1927 onwards, and initially, the leading arm was a trailing arm.

Fig. 11-9. Dubonnet suspension for the Fiat 1500 of 1935 made for a neat installation, using a fork-ended, tubular front cross-member to carry the king-pin.

The Dubonnet suspension system takes its name from Andre Dubonnet, the creator of an aperitif formula that made his fortune (fortified red wine with arrowroot flavoring). He was no engineer; but he was a car enthusiast and racing driver, of no mean stature, whose mind held a great deal of technical understanding. To work on his technical ideas, he hired an engineer named Gustave Chedru and set him to work in a small factory at Neuilly, a western suburb of Paris. Armed with a demonstrator car based on the 1929 Hispano-Suiza, Dubonnet and Cpedru went to Detroit to try to sell licenses for the use of the invention and, in contrast to a multitude of hopeful inventors who have tried the same thing, they succeeded. Their success was partially because of supply problems with the coil springs needed for GM's own system. But General Motors acquired a Dubonnet license, redesigned the system with much simplification, and placed it in mass production for the 1934 Chevrolet and Pontiac. The spring cylinders and hydraulic dampers were manufactured by Delco-Lovejoy. But is was not fully developed and was not considered successful. After two years, Dubonnet suspension disappeared from Chevrolet and Pontiac. But in 1936 it reappeared on the Opel Super Six, and in 1937 on the Vauxhall Ten and Twelve. Both Opel and Vauxhall were parts of General Motors.

The last GM production cars to use Dubonnet suspension were the Vauxhall Wyvern and Velox from 1948 to 1951 (Fig. 11-8). The system was different from earlier versions in several ways, however. The suspension cylinders were placed horizontally and mounted on king-pins and steering yokes at opposite ends of a tubular cross-member. The rear end of the cylinders had a pivot shaft with a leading arm that carried the wheel hub. This pivot shaft was part of a short transverse torsion bar, and the cylinder contained a coil spring in compression to modify the action of the torsion bar. The rear cover of the cylinder contained a piston-type hydraulic damper.

Always keeping a keen eye on GM's technical activities, Fiat embraced the Dubonnet suspension after seeing the 1934 Chevrolet and adapted the Dubonnet front end to its six-cylinder 1500 series for 1935-a design that was produced with minor changes only up to 1949 (Fig. 11-9).

Alfa Romeo's Bimotore, the monstrous twin-engined single-seater built in 1935 for the Scderia Ferrari, also had Dubonnet front suspension. The last car to use Dubonnet suspension was the Delahaya 175 built from 1947 to 1951.

Chapter 12
Evolution of Lateral A-Arms

Most of the independent front suspension systems used on cars from the early Thirties to date belong in the lateral A-arm class (which in recent times has been challenged by the MacPherson type of suspension—a challenge that is not only present, but gaining strength).

Yet, it was not until the 1930's that mass-produced cars adopted any system of front suspension. Although the lateral A-arm type of design had a long history, starting with the 1878 Bolle steamer (as described earlier), it was forgotten until French automotive historian Baudry de Saunier rediscovered Bollee's remarkable precocity about 1935. No doubt the engineers who pioneered the use of independent front suspension lived in ignorance of Bollee's work and reasoned their way through to their various designs by their own technical knowledge and logic. We find traces of their activity on both sides of the Atlantic.

Wolseley engineers A. A. Remington and A. J. Rowledge were experimenting with two different types of independent front suspension as early as a 1911. The first system had equal-length parallel A-arms with a transverse leaf spring attached to the lower A-arms on each side. In the second system, the leaf spring replaced the lower A-arms, creating unequal-length geometry. But neither got into production, Wolseley contentedly continuing with beam axles and semi-elliptic longitudinal leaf springs into the post-1945 era.

Sizaire, Julian and Peugeot

Resumption of car production after World War I led to a rush of new front suspension designs, some of them so advanced that they reveal a deeper understanding of the duties of the wheels than they have been given credit for.

In 1922 our old friend, Maurice Sizaire, now united with his brother Georges in building cars with the Sizaire Freres trademark in a Paris suburb, created a fine luxury car with all-independent suspension, using similar layouts front and rear. He used triangular upper control arms and a low-mounted transverse leaf spring acting as lower control arms. This gave short-and long-arm geometry, useful for restricting variations in track and counteracting roll camber. But the geometry was variable because the pivot axis of the lower arms was undefined and moved back and forth according to magnitude of spring deflection.

Later, Sizaire patented a system with the spring acting as the upper control arms, and lower arms that were so long they overlapped. To avoid interference between the lower arms, they were made asymmetrical. The patent drawings show that Sizaire had clearly anticipated the twin I-beam Principle.

Two American cars of 1921, Parenti and Adria, had front suspension systems with equal-length control arms and dual transverse leaf springs, but neither got beyond the prototype stage.

The 1922 Julian, designed by Julian S. Brown of Syracuse, New York, was produced in small numbers for about three years. It had triangulated lower control armms pivoted on the tubular backbone frame and an upper transverse leaf spring. It was link an inverted Sizaire design with the same advantages and weaknesses.

A strikingly similar design was developed by Georges Broulhiet for Peugeot in 1931 (Fig. 12-1) and introduced on the 201. Broulhiet was an

Fig. 12-1. Half-axles pivoted in the middle became lower control arms, while a transverse leaf spring served as the upper arms on the Peugeot 201 in 1931.

Fig. 12-2. Independent front suspension with equal-length A-arms and transverse leaf spring appeared on the San Guisto in 1923.

independent expert whose contributions towards complete understanding of vehicle dynamics are described in Chapter 25.

San Giusto, Ceirano and Alvis

An Italian engineer, Guido Ucelli, was working on the equal-length A-arm idea when the owner of the San Giusto works, Cesare Beltrame, commissioned him to design an advanced light car. The 1922 San Giusto had fabricated narrow-base A-arms of equal length in both front and rear suspensions, with a spring above the upper control arms. The San Giusto was produced in very small numbers only (See Fig. 12-2).

About the same time another Italian engineer, G. Parisi, patented a more radical design which was used on the Ceirano S-150 between 1927 and 1929. Instead of a transverse leaf spring, Parisi used coils mounted horizontally, inboard of the control arm pivot shafts, and actuated via a crank arrangement.

Captain G. T. Smith-Clarke, chief engineer of Alvis, developed a design with quadruple transverse leaf springs for a front-wheel-drive design in 1926. Because the outer ends were linked to a common king-pin, the parallelogram actually functioned as a system of equal-length upperr and lower triangular arms. Because of the horizontal spacing of the springs, the Alvis design was able to take up great thrust forces without upsetting the front wheel alignment or geometry.

Rohr, Mercedes-Benz and Tatra

Other engineers, faithful to rear wheel drive, were experimenting with simpler layouts. Hans Gustav Rohr, in 1919-20, built a prototype on which the upper and lower ends of the king-pins were attached to a pair of transverse leaf springs, one above the other. No other locating members were used, and Rohr remained faithful to this type of construction even on the big 8-cylinder Rohr that was in production from 1928 to 1931.

The same idea, with some safeguards, was taken up by Hans Nibel, chief engineer of Mercedes-Benz during the period from 1929 to 1934. In 1931, the company introduced Type 170 (Fig. 12-3) with one upper and one lower transverse leaf spring in the front suspension plus a central link for shock absorber attachment and wheel location in the event of spring fracture. This placed severe torsional stresses across the spring leaves during braking, and Nibel soon abandoned the design.

But Nibel developed his ideas of both wheel location and geometry in designing the 380 of 1932, doing away with leaf springs altogether and fitting unequal-length upper and lower triangular arms and vertical coil springs. An intermediate step was brought out the following year-though the design was certainly made prior to the 380-on the 290 which had a lower transverse leaf spring and upper triangular arms extending beyond the pivot shaft to work on two inboard coil springs in a rocker-arm type of action (Fig. 12-4).

Hans Ledwinka, the brilliant chief engineer of Tatra, in designing the Type 77 in 1933, adopted the dual transverse leaf spring layout, using the springs themselves as lateral control arms without the benefit of triangulation. Apparently the springs had the strength to withstand the braking thrust, for the same system was retained for the Type 87 in 1937 (Fig. 12-5), which was built until 1940.

BMW, Peugeot and Fiat

Most chassis engineers of the time did refuse to put such cross-bending loads into transverse leaf springs, preferring to add radius arms in one form or another.

Max Friz and Fritz Fiedler of BMW starting with the Dixi, made under Austin license as a base, gradually developed more progressive

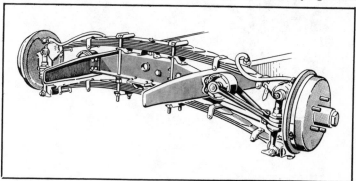

Fig. 12-3. Upper and lower transverse leaf springs were used on the Mercedes-Benz 170 in 1931, with a vane-type shock absorber linked to the king-pin.

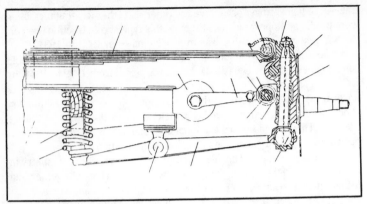

Fig. 12-4. Mercedes-Benz patent from 1933 reveals upper control arm working inboard coil spring, plus lower leaf spring and shock absorber.

chassis designs and brought out a light car with independent front suspension in 1932. The front wheels were attached to long, robust radius arms and a transverse leaf spring-a simple, but primitive, design which involved large camber changes and improper steering geometry. But it led BMW to more satisfactory leaf spring suspension by 1935.

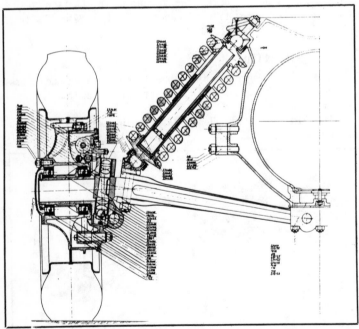

Fig. 12-5. Front swing axle was used on Tatra 24 truck chassis in 1931. Spring unit with shock absorber provided no guidance, and king-pin angle was locked into the half-shaft.

Fig. 12-6. Front-wheel drive Audi of 1937 had independent front suspension with a lower transverse leaf spring and upper A-arms. Rear end had an I-beam axle and transverse leaf spring.

With the 321, 327 and 328 family, Fiedler went to lower A-arms and an upper transverse leaf spring. Taken to England in 1945, he copied his own design for the new Frazer-Nash and Bristol cars appearing in 1946.

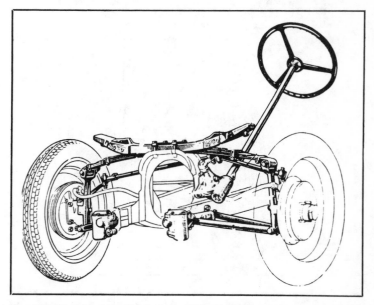

Fig. 12-7. Fiat 500 of 1935 had unequal-length upper and lower A-arms plus an upper transverse leaf spring. Hydraulic dampers were linked to lower control arms.

French manufacturers preferred the lower transverse leaf spring, as pioneered by Peugeot in 1931. Delage began using it in 1932; Berliet, Donnet and Talbot in 1933; and Delahaye in 1934.

Early front-wheel-drive cars, such as the 1931 DKW and 1932 Adler, had used the upper transverse leaf spring as a stress-carrying locating member in combination with triangulated lower control arms. The same layout was adopted for the 1935 Audi (Fig. 12-6). And Fiat's versatile engineering genius, Dante Giacosa, used it for the sensational rear-axle drive 500 (Topolino) minicar of 1935 (Fig. 12-7), which served as inspiration for John Cooper when building the suspension systems for his first 500-cc Formula 3 racing car in 1946. (See also Fig. 12-8).

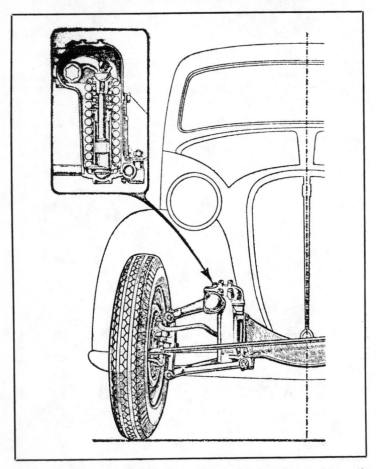

Fig. 12-8. In 1936 Fiat used coil springs enclosed with the hydraulic dampers in oil-filled cylinders, worked by a rocker-arm from the upper control arm on the 508 C Balilla.

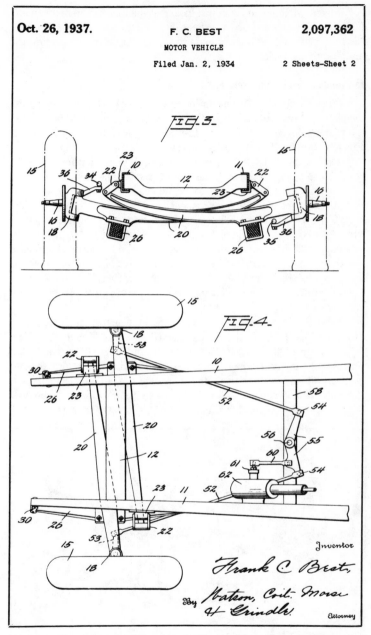

Fig. 12-9. Frank Best's patent (assigned to Packard) for twin I-beam suspension, issued in 1937. Best intended it for semi-elliptic leaf springs and an odd, retracted steering linkage.

Eugene Mathieu no doubt had looked at the Sizaire patent when he designed the twin I-beam front suspension with a transverse leaf spring for the 1931 Unic. But shortly afterwards, Unic ceased car production to concentrate on trucks.

Variations of Short and Long Arms

Frank C. Best, experimental engineer at Packard, found a way to combine the twin I-beam layout with longitudinal semi-elliptics. He paterted his design in 1934 (Fig. 12-9), but Packard never used it on a production model. Twin I-beam suspension was left in oblivion until James Heywood, with G. H. Muller and M. L. Jurosek, revived it for the 1965 Ford pickup truck range using coil springs and leading arms.

A possible predecessor for some of the design elements in the Ford system can be found in the front suspension of the radical North-Lucas of 1922. This was the idea of Oliver D. North, formerly of Straker-Squire, who designed the car for Ralph Lucas. Of utter simplicity, it had swinging half-axles with diagonal leading arms and enclosed vertical coil springs. But only two or three cars were built. The North-Lucas really was a predecessor of the front swing axles as adopted by Hans Ledwinka for the Tatra 24 truck chassis of 1931 and by Michael Parkes for the Hillman Imp of 1963.

One of the weirdest variations on the SLA theme that actually got into production was probably that used on the Glas Isar T-700 and other Glas cars between 1963 and 1969. The lower control arm was a normal A-arm linked to the lower ball joint (Fig. 12-10). It provided a platform for the certical coil spring which was retained in a low tower that formed part of the body shell. The upper ball joint was located above tire level and was connected to the front end of a semi-leading arm pivoted from a bracket in

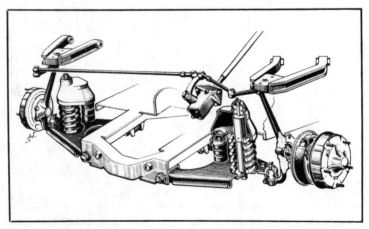

Fig. 12-10. Glas Isar combined a lower lateral A-arm with an upper triangular leading arm. Steering linkage worked without an idler arm.

Fig. 12-11. Front-wheel drive Citroen 7 CV of 1934 was first to use lateral A-arms and torsion bars. This design was ahead of its time in many ways.

the body shell. This gave a geometry that involved large changes in caster angle during deflection-and opposite caster changes in roll. Camber changes, however, were quite small and the roll center was at a high level. The T-700 was a minicar, and its front end was acceptable for handling precision and cornering stability within its speed range.

Rover used something similar on the 2000, introduced in 1963, though the leading arm had considerable length and thereby minimized caster angle variations, having its pivot shaft anchored in a bracket on the cowl structure. The coil spring was removed from the lower control arm and placed in a near-horizontal position in the upper fender panel, working against the cowl. Near the forward end of the leading arm was an angled abutment that compressed the coil in jounce travel. With this design, the Rover obtained an exceptionally high front roll center seven inches above ground level and suitably low spring rates. The system was in production for the former Rover 3500 up to 1976.

Slow Adaption of Citroen's Torsion Bars

The merits of Nibel's design for the Mercedes-Benz 380 of 1932 were immediately recognized by chassis engineers both in Europe and America and it became the most copied of all front suspension systems. It took much longer, for the Citroen 7 CV of 1934-an equally advanced concept, to win recognition and breed a school of imitators. This design was the work of Maurice A. Julien, a scientifically-minded chassis expert who had been educated as a mining engineer. His design for the 7 CV Traction Avant (Figs. 12-11 and 12-12) possessed upper and lower triangulated control arms of unequal length, with longitudinal torsion bars extending backwards from the lower A-arm pivot shaft to an anchorage point in the cowl structure.

201

Fig. 12-12. Detail of the front suspension on the 7 CV Citroen shows complicated steering linkage, forged-steel suspension arms, and friction-type shock absorber.

Fig. 12-13. Drag strut anchored in rubber bushings at its front end was first used on the 1948 Morris Minor. The strut served to triangulate the lower control arm, which worked a longitudinal torsion bar.

First to copy this layout was Ernesto Maserati, who used it on the new 8 CL monoposto of 1937. Harry Rush Produced a similar design for the 1946 Riley 1-1/2 and 2-1/2 Liter models, which Gerald Palmer modified for the Riley Pathfinder of 1954 by replacing the lower A-arm with a transverse I-arm and diagonal compression strut.

Alec Issigonis, who had joined Morris in 1936 as a suspension expert after six years with Humber, drew up a Citroen-type torsion-bar suspension for the new Morris Ten, but it was discarded in favor of an I-beam axle and semi-elliptic leaf-spring design by Hubert N. Charles-the same Charles who had used torsion-bar suspension on his R-Type MG in 1935 (though with equal-length lateral A-arms). Issigonis developed a new torsion-bar setup for the 1949 Morris Minor. This design was a trend-setting one for it was the first to eliminate the king-pin and carry the steering knuckle on ball joints. A less advanced system, probably derived from the Rileys', was installed on the 1948-model six-cylinder Morris and Wolseley crrs (Fig. 12-13).

Alfred Boning adopted a similar design for the BMW 502, which went into production in 1951 and formed the basis for the 503 and 507 sports models. Then, beginning with the 1959 models, Chrysler shed its coil springs in favor of torsion bars. This system is still in use on many Chrysler, Dodge and Plymouth models. The Aspen/Volare family is noteworthy for using transverse torsion bars in combination with lateral A-arms. While the French-built Horizon has the torsion bar suspension from the Simca, the U.S.-built Omni/Horizon models have VW-type Mac-Pherson front suspension systems.

Torsion bars have been used in cars since shortly after the turn of the century, originally as part of the wheel-locating linkage. George Johnston of Arrol-Johnston in Scotland and Eugene Mathieu of Excelsior in Belgium were experimenting with torsion bars to restrict body roll as early as 1904.

In 1911, a Wolseley engineer, A. A. Remington, designed a rear suspension system with torque tube drive and cantilever springs whose center brackets were formed as sleeves for the ends of a transverse torsion bar intended to restrict body roll. It was patented, but apparently not used in production.

The best known anti-roll device in Europe in the Twenties was the Adex stabilizer, designed by Albert de Coninck of Excelsior. It was a multi-rod system which helped locate the rear axle laterally, as well as restricting body roll. As used on the 1921 Excelsior, it consisted of two transverse rods (torsion bars) running from the top of the differential casing to shackled brackets on the frame side members. The left shackle also carried a rocker arm which had a separate rod connected directly to the opposite shackle.

The first to use the double-elbow stabilizer bar as it exists today was J. G. Parry Thomas, who installed them at both front and rear ends of the magnificent 1921 Leyland Eight. Incidentally, the earliest patent for torsion bar suspension is dated 1878 and was taken out in Norway by Anton Lovstad who envisaged its use on horse-drawn carriages.

Gregoire and Variable-Rate Springs

Both torsion bars and coil springs have *constant-rate action*, unless diameter or coil spacings are different in certain areas of the spring. Laminated multi-leaf or tapered single-leaf springs, on the other hand, have *variable-rate action*. This is valuable in cars, especially those with great seating and luggage capacity, giving the widest difference between unladen and full-load weights. But leaf springs introduce other problems, and many engineers have tried to obtain variable rates with torsion bars and coil springs.

One man who has devoted years of his life to this end, and become known as the leading expert on spring progressivity, is Jean-Albert Gregoire. He was famous as a pioneer of front-wheel drive and promoter of the Tracta constant-velocity joint before he turned his attention to variable-rate springs. In collaboration with Charles Rivolier, he explored the entire field for vehicles of all sizes and presented a prize-winning paper to the French Academy of Science in 1947.

The principle behind all the Gregoire variable-rate springs lies in angular variation of the suspension linkagem A vertical coil spring attached directly to a horizontal suspension arm will work at a constant rate directly proportional to any increase in load. By placing coil springs in tension, rather than in compression, and mounting the spring diagonally in a parallelogram linkage, progressive action is obtained because the spring deflections increase, not proportionally, with additional wheel travel. If initial suspension movement of one inch stretchs the spring one-half inch, an inch-bump occurring towards the extreme limit of wheel travel may correspond to 2-1/2 inches of stretch in the spring. That gives progressively stronger resistance as more and more of the available wheel travel is used up.

Fig. 12-14. Horizontal coil spring working in tension gave progressive-rate action on the Gregoire sports car. Outboard end of the spring was attached to the fabricated upper control arm, and the inboard end to a bracket on the frame.

The 1955-57 Gregoire Sport was suspended entirely by variable-rate coil springs (Fig. 12-14). At the front end, a near-horizontal coil spring was mounted between the lateral A-arms, its upper end anchored at the top of the king-pin and its lower end linked to a bracket on the chassis. In all jounce movements, spring travel increasingly outran wheel travel. There was not much rebound in the design, for at static height both control arms were angled sharply down towards the wheel hub, and clearance problems for the upper arm put a limit on rebound travel.

Gregoire also sold licenses to other manufacturers for use of his variable-rate springs as progressive-action load levelers. Subsequent work in this area has usually gone that route, though there is one interesting exception: The 1957 Studebaker had variable-rate coil springs in the independent front suspension, due to uneven spacing between the coils. However, the overall level of chassis construction at Studebaker was such that the value of this refinerent was lost in practice.

Chapter 13
American Developments

For the longest time, a lot of people in the American auto industry thought that ride, handling and control problems would go away if we only got good roads with smooth surfaces. Eventually it became clear that this would not be, the case, and that the problems would always persist unless something was done about springs, axles, and such things. It was General Motors that led the way. Its pioneering efforts began towards the end of the 1930, after Murice Olley had been brought in from the American Rolls-Royce Company (which folded earlier that year) as head of Cadillac's suspension engineering group.

Rolls-Royce in Great Britain was one of the first car companies to set up a full-scale bump rig to examine steering phenomena. This was done about 1925 at the instigation of William A. Robotham who had joined Rolls-Royce as an apprentice in the experimental department in 1919. Olley was, of course, familiar with Robotham's work in his capacity as chief engineer of the Rools-Royce works at Springfield, Mass. Olley was to have tremendous influence on the evolution of usspension systems and our knowledge of automobile dynamics in the years to come. It was he who coined the terms oversteer and understeer. Later, he laid down the rules governing weight transfer during cornering, the importantce of roll stiffness and roll-steer effects.

Impact of Olley and GM Goes IFS

Born in England in 1889, Maurice Olley was educated at the University of Manchester and Birmingham Technical College. After graduating, he went to work for a firm of machine tool manufacturers. By 1912 he was their chief tool draftsman, but he left later that year to join Rolls-Royce as draftsman of factory equipment in the Derby works. In July 1917, he was put on the personal design team of Henry Royce. Two years later, he went to America as chief engineer of the chassis division for the American branch of Rolls-Royce.

Independently of Hans Nibel's work in Germany, Maurice Olley reasoned his way through to the unequal-length lateral A-arms and drew up a basic scheme with a vertical coil spring mounted on the lower A-arm and abutting against a bracket on the front cross-member, while the upper A-arm worked a Delco piston-type hydraulic shock absorber (Fig. 13-1).

At the time, General Motors also got interested in the Dubonnet suspension. Olley spent a lot of time adapting it to the Cadillac and developing the system. His group accomplished a fantastic job by getting two completely different systems from the idea stage to the ready-for-production stage in less than two years!

Both suspension systems were tested on full-scale Cadillac chassis on the K2 rig (which was set up in 1932). Both were developed to full satisfaction for the goals of that time. In March 1933, the GM Research Laboratories were ready to demonstrate both systems to the car divisions. The chief engineers of the divisions, the general managers and the top technical management 'at GM took a test ride with two prototype Cadillacs—one with Dubonnet and one with Olley's A-arm design—and a Buick with a normal front axle for comparison. They went from Cadillac's factory in Detroit to Monroe, in southern Michigan, and back, taking turns driving and riding in the three cars.

At the end of the day, all the divisions put in a claim for independent front suspension. F. A. Dutch Bower, chief engineer of Buick, was the first. He picked the lateral arm system for its mechanical simplicity and

Fig. 13-1. Knee-action front end of the 1934 Cadillac had very long lower control arms and short upper ones linked to Delco shock absorbers.

low production cost. Charles L. McCuen, then chief engineer of Oldsmobile, echoed him. For Ernest V. Secholm, chief engineer of Cadillac, standardization of Olley's design was a matter of course. His engineers had developed it, and he was convinced of its superiority.

But what about Chevrolet? Chief engineer James C. Crawford wanted it, and he got the full support of William S. Knudsen, the general manager of Chevrolet. But Ormond E. Hunt, corporate engineering vice president and a former chief engineer of Chevrolet, turned them down. Hunt told Knudsen there were not enough centerless grinding machines available in the whole U.S. to grind the wire for the springs that would be needed to make independent front suspension systems for all Chevrolet cars.

But Knudsen and Crawford did not give up that easily. They saw that the Dubonnet system had a smaller and different coil spring and found a way to make it in large numbers. That's the reason Chevrolet's 1934 model (Fig. 13-3) had Dubonnet suspension, while the senior makes used the Olley system. Pontiac was allowed to share the Chevrolet version of the Dubonnet system so that all 1934 GM cars offered independent front suspension in some form (Fig. 13-2).

Field experience soon proved that the Dubonnet suspension was unripe for mass production. It was unreliable in operation, difficult to service and repair and unpredictable. It was discarded during 1936, and since nothing had changed in the machine tool industry, Chevrolet and Pontiac were compelled to revert to I-beam front axle suspension systems. It was not until 1939 that Chevrolet and Pontiac got their own versions of the lateral A-arm system. It was standard on Pontiac and optional on Chevrolet.

Chrysler, too, with IFS

It may be surprising to some readers to learn that Chrysler wasn't first in the U.S. with independent front suspension. Chrysler's great Zeder, Breer and Skelton engineering team was justly famous far beyond automotive circles. But, actually, it's a small miracle that Chrysler managed to get independent front suspension into its 1934 models at all (Fig. 13-4).

Look at what these men were busy with in the 1932 to 34 period: Fred M. Zeder was engineering vice president, with responsibilities including production engineering. The ultimate responsibility for starting or not starting a new suspension study program lay with him-but apparently neither did he see the need for it, nor was it suggested to him. His main research interest at the time is believed to have been automatic transmissions (Fluidrive). Owen R. Skelton was making investigations into high-compression cylinder heads and hemispherical combustion chambers. Carl Breer was up to his ears in the Airflow-and, revolutionary as it may have been in other ways, it did not include independent front suspension.

Somehow Chrysler found out in the nick of time what GM's plans were, and Zeder gave instructions that an independent front suspension

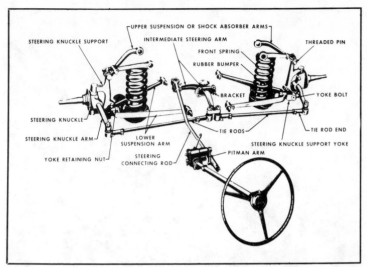

Fig. 13-2. Each front wheel had its own separate steering linkage up to the intermediate steering arm on the 1934 Cadillac. Suspension system had a low roll center, and no stabilizer bar was used.

system should be ready for the 1934 Chrysler Six, Dodge, De Soto and Plymouth (all but the Airflows). The engineers who were saddled with that responsibility were A. Griswold Herreshoff and Harry T. Woolson.

Herreshoff had joined Chrysler in 1927 and was assistant chief engineer of the passenger car division of Chrysler Corporation in 1932. He

Fig. 13-3. The inner workings of the Dubonnet suspension of the 1934 Chevrolet took place in this oil-filled cylinder. The same crank that compresses the coil spring also commands the shock absorber.

209

Fig. 13-4. Chrysler's first independent front suspension bore a close resemblance to Cadillac's but Chrysler engineers had chosen a simpler steering linkage.

was born in Bristol, Rhode Island, into a family famous for many generations as yacht builders. His own interests lay on land, however. He was a graduate engineer from MIT and a former Mack Truck engineer. His background also included four years in the Rushmore Laboratories.

Woolson, on the other hand, was a chassis engineer and had been known as a chassis expert since he came to Chrysler (then Maxwell) in 1922. He was born in Passaic, New Jersey, and graduated from the Stevens Institute of Technology in 1897. After many years of non-automotive work, he stepped into the auto industry as truck engineer with Packard in 1915 and then became a research engineer for Studebaker. Here he met Zeder, Breer and Skelton—a meeting that settled the rest of his career.

The Chrysler design looked like a Chinese copy of the Cadillac system and worked satisfactorily. Herreshoff was promoted to director of research for Chrysler Corp. in 1938, and Woolson was named executive engineer for Chrysler Corp. in 1935 (Fig. 13-5).

Studebaker, Late but Good IFS

All of the independent auto companies, including Studebaker, were unprepared for GM's sudden step to independent front suspension. It took the old-established company from South Bend two years to catch up, introducing its own first system as an option on the 1936 Dictator (Fig. 13-6). To the credit of its creators, it had very little in common with other U.S. designs, going back to the Peugeot 201 version for its inspiration.

Studebaker borrowed some of the geometry, perhaps, from the Cadillac system, for it used a long lower control arm and a short upper one. Instead of coils, Studebaker used a transverse leaf spring clamped down to the front cross-member at the center and shackled to the outer ends of the lower A-arms at the ends. Hydraulic vane-type shock absorbers were mounted on the frame side rails just below the upper A-arms and linked to the steering knuckles.

This design worked very well in practice. The control arms were braced to take brake thrust, and the bulk of the leaf spring was carried as sprung weight. Now, why had not this elegant solution occurred to the Chevrolet and Pontiac engineers when they were told of the shortage of coil spring grinders. this seems incredible since they were tooled-up to make leaf springs and had always been using them to locate the front axle.

The Studebaker suspension was not produced hurriedly, and the idea stemmed from none less that the chief engineer, Barney Roos. A New Englander by birth, he graduated from Cornell in 1911 and began his career with General Electric. His automobile experience included years as chief engineer of such famous makes as Locomobile, Pierce-Arrow and Marmon. He came to Studebaker in 1926 as assistant chief engineer (to Guy P. Henry), becoming chief engineer in 1928. Roos redirected Studebaker's entire engineering effort with great success, giving priority to engine and

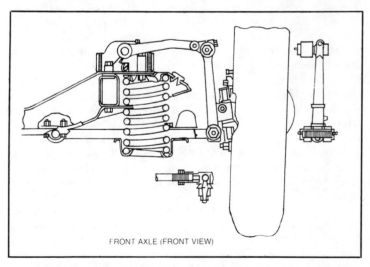

FRONT AXLE (FRONT VIEW)

Fig. 13-5. Front view of Chrysler's 1936-model front end shows high knuckle arm , from king-pin (to keep roll center from going underground) and spring well in lower A-arm to provide adequate travel.

drive train and leaving chassis design for the second round. That's why Studebaker was late in getting independent front suspension into production. When it came, it was in no way inferior. Still, when a new engineering team headed by Roy E. Cole and William S. James designed the Studebaker Champion compact car (which came out as a 1939 model) they gave it coil springs and lateral A-arms.

Later and Less: Packard, Hudson and Nash

After thoroughly reviewing all possible alternatives to the I-beam axle, Packard also ended up with a system using lateral A-arms. Packard's work in this area was mainly the responsibility of a wonderfully inventive engineer named Forest McFarland who had joined Packard as a draftsman in 1924.

Late in 1927, McFarland was assigned to the chassis group under Walter Griswold, and in 1933 McFarland and his associate, Henry Krebs, began to study independent front suspensions. Chief engineer Cylde R. Paton finally narrowed the choice down to a few variations on the coil spring and SLA theme.

Packard's design used two A-arms of unequal length-the lower one a two-piece assembly that was about twice as long as the upper one. A vertical coil spring was mounted on the leading lower arm working against the base for the upper-arm pivot shaft, which also actuated a vane-type hydraulic shock absorber. The trailing member of the lower arm went back to a ball-mount on the frame and could take enormous fore-and-aft loads. The system went into production late in 1935 on the 1936-model Packard 120. By 1937 it was extended to the Packard Eight and Super Eight models and it remained in use through 1940. Then a new version with improved geometry, beefed-up components, and lower manufacturing cost, was introduced for the Clipper and Super Clipper.

The 1935 models of Hudson and Terraplane, Nash and Lafayette, were available with a semi-independent front suspension known as Axleflex. It was invented by J. A. Baker, a former Willys-Overland engineer who had gone to Hudson in 1933. It consisted of a solid axle cut in two, with a central parallelogram. It was, therefore, not truly independent, as a one-wheel bump would cause a camber change on the opposite wheel. Its standard semi-elliptic leaf springs were burdened with brake thrust and other horizontal forces in the system. It also had the disadvantage of excessiye roll understeer.

Clearly, Axleflex was only a stopgap solution; but it took both Nash and Hudson until 1940 to develop their own systems and get them into production. Millard H. Toncracy at Hudson chose unequal-length lateral A-arms with slightly diagonal pivot axes (to control toe-in during deflections) and coil springs. Meade F. Moore at Nash pioneered a novel version of the sliding pillar with a coil spring mounted on top of the oversize king-pin.

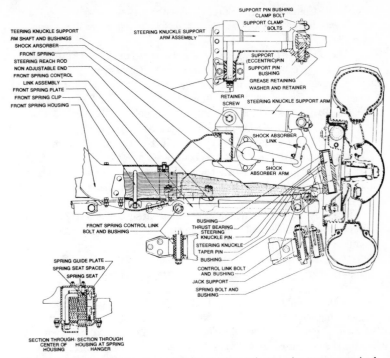

Fig. 13-6. Studebaker's independent front suspension used a transverse leaf spring, in combination with unequal-length upper and lower A-arms on the 1936 Dictator and President.

Live Henry, Live Axle

Ford was left far behind its competitors in chassis design. It was Henry Ford himself who insisted on continuing the Model T's front axle and cross-springs for the Model A in 1927, and it was Henry Ford who stifled every attempt to modernize the the suspension systems for the V-8-powered cars. Forward-looking engineers like Laurence Sheldrick and Eugene Farkas wanted to keep up with the competition, but were frustrated by Henry Ford's determination to black-ball all such proposals. When the Lincoln-Zephyr was being designed, independent front suspension was in consideration; but the production model used the same cross-spring arrangement as the standard Ford. There was even less of an attempt made to get a modern front end on the first Mercury, since it shared all basic components with the Ford.

During the war years, Sheldrick and his men did quite a bit of work on independent front suspension in preparation for postwar models. But the old cross springs were not abandoned until Henry Ford was gone and a new engineering staff, under ex-Oldsmobile chief engineer Harold T. Youngren, had taken over at Dearborn.

The 1949 model was the first Ford with independent front suspension. It was designed by H. S. Currier and closely resembled the contemporary GM Layout.

In his stubborn resistance to replace the front axle, Henry Ford as not isolated, however. Smaller companies such as Graham-Paige and Hupmobile, never came close to using independent front suspension on a production car. Nor did Franklin, Auburn, Willys, Reo or Pierce-Arrow.

The 1936 Cord had independent front suspension, but that was because it had front-wheel drive, and a live front axle would not have allowed the low coffin-nose styling. The 1936 Cord had trailing arms, one on each side, with a wide base to provide adequate earing area for the pivot shaft. A transverse leaf spring was located below and attached via vertical rods to the mid-points of the trailing arms, so that spring deflection was circa one-half of wheel travel.

Regardless of the fact that independent front suspension originated in Europe, and that the scientific approach taken by such engineers as Hans Nibel of Mercedes-Benz and Georges Broulhiet as a consultant to Peugeot, paved the way for mass-producing reliable and well-behaved systems, there can be no doubt that progress in chassis design was a two-way street.

American influence had a lot to do with the evolution of suspension engineering, both in Britain and on the Continent, though mostly in England. When Rolls-Royce decided to use independent front suspension on the 1936 Phantom III, the venerable Derby firm bought a license from General Motros and built its system under GM patents. The Studebaker system appeared on the 1937 Humer Snipe and Super Snipe and the 1939 Hillman "14." That seems strange until you realize that Barney Roos had hired himself out as an engineering consultant to the Roote United Kingdom. He returned to the States in 1940 to join Willys in Toledo—just in time to go to work on the military truck that was to become the Jeep.

Chapter 14
MacPherson Strut History

What a MacPherson-type suspension is, we have discussed at length in Chapter Seven. The first production models to use it were the British Ford Consul and Zephyr introduced in 1950; but it was invented and patented by General Motors and was a feature of a small car under development at Chevrolet from 1945 to 1947.

Even the prior art contains valid approximations to the same design principles. The 1926 Cottin-Desgouttes had an independent front suspension system which no doubt started from the idea of combining sliding pillars with an underlying transverse leaf spring. The result was a spring leg whose outer cylinder was attached to the chassis frame near its upper end, with an inner rod serving as king-pin and telescoping into the cylinder. The leaf spring made a lower control arm. By stretching the sliding pillar, inclining it inboard from the wheel hub and making the spring leg a locating member, the Cottin-Desgouttes arrived at MacPherson-type geometry (Fig. 14-1).

In 1927, a Fiat engineer named Martinotti received an Italian patent for a new front suspension design with a low, triangulated control arm and inclined spring legs with coil springs wound around their upper portion, containing telescopic-hydraulic shock absorbers. This invention corresponds closely to the MacPherson-type suspensions adopted in later years for certain Fiat models (127 and 128); but it was not developed at the time of its invention, and Fiat stayed away from MacPherson suspension until Ford and others had thoroughly proved its qualities.

There is one significant difference between Martinotti's patent drawings and those in the MacPherson patent. In MacPherson's drawings, there is a front stabilizer bar whose ends are shaped as drag struts and, thereby, form the forward part of a triangular lower control arm. The other part is a simple lateral rod. The spring leg stands on a base extending from the upper portion of the wheel hub and shows very slight inclination, giving a long scrub radius. Naturally, each spring leg housed a telescopic-hydraulic shock absorber. Significantly, the upper ends of the spring legs were mounted in rubber bushings to insulate suspension noise and vibrations from the body structure.

Chevrolet Invention/Ford Production

The patent was issued on January 6, 1953, to Earle Steele MacPherson, who had applied for it as early as March 21, 1947. He had joined Chevrolet in 1935 after a year in the GM Central Engineering Office, and in 1945 he became chief engineer of the Light Car Project, known under the code name Cadet. His former experience included 12 years at Hupmobile, where he was responsible for most of that make's innovations. He was born in Highland Park, Illinois, in 1891 and graduated from the University of Illinois in 1915. He had joined Chalmers Motor Company in Detroit on leaving school and spent a lot of time in Europe during World War I working on aircraft engines for the U.S. Army.

But it was Ford, not Chevrolet, that was first to produce cars with the MacPherson-type suspension. That came about because the Cadet was

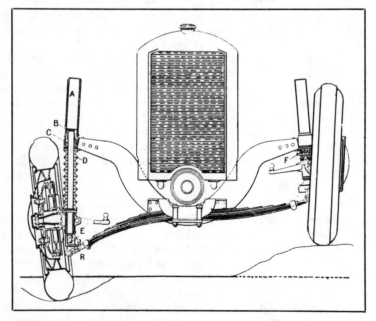

Fig. 14-1. Cottin-Desgouttes in 1927 was approaching MacPherson-type suspension with this front end using guiding struts and transverse leaf spring acting as a lower control arm.

216

Fig. 14-2. Chapman struts of the Lotus Elan of 1964 were actually MacPherson spring legs applied to the rear (driven) wheels. But the Elan did not use MacPherson front suspension, for it had lateral A-arms.

killed, and Earle S. MacPherson left GM and went to Ford in September, 1947. His first task was to direct the engineering of two dramatically new post-war cars for Ford of England. After the success of the Consul and Zephyr, the same front suspension was used for the 1953 Anglia, 1962 Cortina, 1969 Capri and many other British Fords (Fig. 14-2).

MacPherson became engineering vice president of Ford in 1952, and in 1956-57 laid out the Falcon, using an SLA design with coil springs mounted above the upper control arm, and this system was inherited by the Mustang. He retired in 1958 and died in 1960. It was not until 1977 that American-made Fords began using MacPherson-type suspension (Fairmont and Zephyr), adding the Mustang and Capri in 1979.

Peugeot Leads Rush to Follow

The first European non-Ford designers to adopt MacPherson suspension was Marcel Dangauthier, director of design engineering for Peugeot, with the 404 of 1960. All subsequent Peugeot models used it, whether rear- or front-wheel drive, from the tiny 104 to the stately 604. Fritz Fiedler, back at BMW, chose MacPherson suspension for the new 1500 of 1963, starting a tradition at BMW where other engineers have subsequently added refinements of their own.

Japanese engineers discovered MacPrerson suspension in the early Sixties. Honda, which now uses it on the Civic and Accord, was first in production with its N-360 in 1964. Datsun followed with the 1965 Bluebird and Toyo Kogyo, with the 1966 Mazda 1000.

MacPherson suspension continued its progress in Europe to the extent of reaching Volkswagen, beginning in 1968 with the 411. The Beetle shed its trailing links in favor of a MacPherson design in 1971 and began a total switch with the Passat in 1973, adding the Golf and Scirocco in 1974. The Rootes Group began using MacPherson suspension with its Hielmen Hs in 1966, adding the Avenger in 1970. Even Volvo, after years of objecting to the system's sensitivity to errors in wheel alignment, adopted MacPherson for the 240 and 260 series in 1976.

Paradoxically, it seems that General Motors, where MacPherson executed the original design, has been among the slowest in its adoption. The first GM car to use MacPherson suspension was the Opel Rekord in 1977, followed by the Opel Senator and Monza in 1978. The first U.S.-mades to use it were the 1980-model X-cars (Chevrolet Citation, Pontiac Phoenix, Oldsmobile Cutlass and Buick Skylark). Finally, the story has completed the full circle.

Chapter 15
Handling Live Rear Axles

With live rear axles, the suspension system must cope with torque reactions as well as the driving thrust. This posed problems that many pioneer car builders found insurmountable. Some such as Henry Ford with his 1896 quadricycle; chose to dispense with springs altogether. Others chose to fit an unsprung sub-frame that carried the body on springs. This layout was adapted by Locomobile in 1899, and Knox and White in 1900. The 1898 Autocar and 1899 Packard also had unsprung frames. Most of these frames were highly flexible, however, and could almost be regarded as a primitive approach to torsion-bar springing. Long before these American cars were created, European engineers had actually come to grips with the live rear axle.

When Carl Benz built his tricycle in 1885, he placed the springs not between the frame and the body, but between the frame and the driving rear axle. The body was then bolted solidly to the frame, and the first true rear-axle suspension had been created. The engine was mounted in the frame, with chain drive to sprockets on the rear axle. Near each wheel hub a full-elliptic spring was attached to the axle with its upper part clamped to the tubular frame.

Benz fully realized, even from his little single-cylinder which had a maximum output of 2⅔ (yes, two-thirds) horsepower at 250 rpm, that this sort of suspension could not cope with the thrust load. The axle would travel forward, the springs would bend, the driving chains would slacken and jump off their sprockets, and the car would not run. Benz, therefore, fitted radius rods between the lower spring clamp (on the axle) and the engine cradle which was bolted to the frame.

These radius rods were pivoted at the front ends. Their sole, but vital, function was to keep the chain sprockets at a constant distance from each other. They were called *radius rods*, because they formed a radius between the fixed-position driving sprockets and the driven sProckets which moved in an arc (centered on the pivot points) during spring deflections.

Radius Rods for Chain- and Shaft-Driven Cars

Leading constructors of chain-driven cars adopted the Benz-type radius rods and kept them, even when they changed from full-elliptic to semi-elliptic springs. Noteworthy examples are the first front-engine car, Emile Levassor's 1894 Panhard, and the car that set the pattern of design for a decade, Wilhelm Maybach's 1901 Mercedes. The curved-dash Olds, the 1903 Ford, Cadillac and Buick also used chain drive and radius rods, though the designers' choices of springs varied greatly.

In the meantime, Louis Renault had invented shaft drive and used it on his 1898 prototype which carried the rear axle on full-elliptic springs. He used a single radius rod, linking the lowest point on the differential casing with a frame cross-member. About 1902, Sydney Straker introduced a similar setup on the Daimler (built in Coventry). Another early shaft-driven car was the 1899 Adler, whose chief engineer, Franz Starkloph, placed one radius rod above each semi-elliptic spring. When Charles Schmidt, who had worked for Mors in France, joined Packard in 1902, he designed a new model with shaft drive, full-elliptic springs, and long radius rods mounted in a near-horizontal position. In 1903, Louis S. Clarke copied the Renault arrangement on his Autocar, and Louis P. Mooers did the same on the Peerless.

On the whole, however, the industry was slow to discard chain drive in favor of shaft drive, in Europe as well as in America. Fiat used chain drive as late as 1914 (on Tipo 6). The 1912 90-hp Mercedes had chain drive, and chain drive was used on the 1908-10 Peugeot Type 120. The 1909 models of Knox, Reo and Thomas Flyer were chain-driven. International High-Wheelers used chain drive up to 1913 and Mack Trucks still used chain drive in 1926. Oddities such as the 1919 GN and its licensed copy, the 1921 Salmson, had chain drive; their offspring, the Frazer-Nash sports car, used chain drive until 1934.

Mathieu and Torque-Tube Drive

But the road to progress went the way of shaft drive. While some engineers were content to keep the same suspension systems, with leaf springs and radius rods, others searched for better or simpler arrangements.

Radius arms were not the ideal thing for controlling torque reactions in the axle housing, for instance. One of the first to realize this was a French engineer named Eugene Mathieu. He was the sort of engineer it is tempting to classify as a genius, but that's such a tired cliche, as well as being non-descriptive. Mathieu was a methodical technician, a clear and precise thinker, never to be swayed by vague theories. Born in Colmar in 1869, he first studied architecture, but then turned to mechanical engineering. When a friend told him about his defective sewing machine, Mathieu, who had never looked at sewing machines before, calmly redesigned it. This feat led him to a job at the Paris firm of Hurtu, Hautin & Diligeon, makers of sewing machines, bicycles and automobiles.

In the recent years 1901 to 1903, Mathieu's ideas of improvements in automobile engineering brought him to take out a number of patents which included a chassis with torque-tube shaft, bolted to the rear axle and, at its front end, anchored in one large or two small ball-joints, free to swing up and down. If two ball-joints were used, the torque tube had a forked front end. If only one joint was used, it had to be large enough to allow the propeller shaft to pass through its center. The torque tube effectively transmitted the driving thrust into the ball-joints which were usually mounted on a frame cross-member. It also resisted torque reactions in the axle. Mathieu's designs show the axle to be located laterally by semi-elliptic leaf springs which also carried the load. In 1902, he was head of the engineering department for Excelsior at Saventham and produced a model with torque-tube drive. In 1905 he sold a small-car design with torque-tube drive to Minerva.

In the meantime, American engineers were working towards the same solution. As early as 1904, C. Harold Wills had incorporated torque-tube drive on a Ford design (which was not built). It was to be a four-cylinder car, and the rear axle was held in place by semi-elliptic leaf springs with diagonal stays. Either Wills had learned of Mathieu's patents, or he had invented the same thing independently.

Wills and Ford thought enough of the torque tube to adopt it for the Model N in 1906, and it proved so successful that it was made part of the Model T specification in 1908. It was retained for the Model A in 1927, the V-8 in 1932, and all subsequent Fords until 1949. In Ford's layout, a transverse leaf spring was used with diagonal struts to help locate the axle.

In Europe it was Fiat that popularized the torque tube. Its torque tube system, with semi-elliptic leaf springs in the Mathieu tradition, was designed by Cesare Momo in 1905. Until that time, all Fiats used chain drive. The German Oryx went to torque-tube drive in 1908, Mercedes in 1909 and Benz in 1911. In the U.S., Cadillac and Buick were quick to follow Ford's example, and they were far from the only ones. In the years 1910 to 1914, torque tube drive was used by Cutting, Great Western, Havers, Overland, Rambler, Regal and Stoddard-Dayton.

Buick had adopted the torque tube in 1907, and continued it through 1964 on the LeSabre. Cadillac used torque tubes on the Model D from 1905 and the 30 from 1909, but in 1914 went to an open propeller shaft with semi-elliptic springs and a long radius rod on the left side of the differential. Four years later, Cadillac adopted Hotchkiss drive.

Hotchkiss Drive Simplicity

We have seen what Hotchkiss drive is (Fig. 15-1), but where did it come from and why is it called that? It was made in Paris by a factory

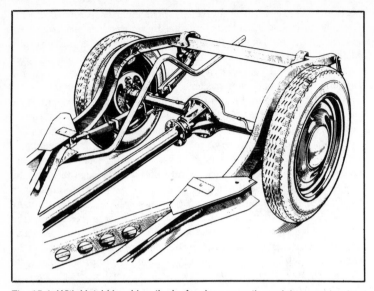

Fig. 15-1. With Hotchkiss drive, the leaf springs carry the weight, provide wheel travel, and locate the axle. This example is a Fiat 500.

established in 1867 by an American manufacturer of guns and munitions, B. B. Hotchkiss. An automboile department was added in 1903, and the chief engineer of Mors, Georges Terrasse, was lured away to design Hotchkiss cars. The first models used Panhard-style chain drive, but Terrasse devised a shaft-drive system without radius rods for the 1905 Model H. He adopted it for a number of Hotchkiss models built up to 1914.

For years nobody copies the Hotchkiss; its theoretical flaws were all too apparent. Then Ernest Henri, a Swiss engineering consultant, was engaged to design a racing car for Peugeot and, despite the high torque of the 7.6-liter twin-cam engine, he chose Hotchkiss drive for the chassis. It was a daring move, and it worked.

European engineers still believed in the torque tube or, at least, hefty radius rods for high-powered cars; but in the U.S., several leading engineers became interested in the simplicity of the Hotchkiss drive system. Among the first was Henry M. Crane, who used it on his 1916 Crane-Simplex. Another was Frank R. Fageol who adopted it for his aircraft-engine-powered 1917 Fageol.

And then came Cadillac in 1918. Of course, Henry M. Leland and Ernest E. Sweet, who left Cadillac in 1917 to start Lincoln, wanted no part of Hotchkiss drive. But a young man named Benjamin H. Anibal, who had come to Cadillac from Oldsmobile in 1911 and worked his way up to the position of chief engineer by 1918, approved it. He was against the torque tube because of its high unsprung weight, and Cadillac dropped it in 1913. After that, it was a relatively small step to do away with the radius rods and let the springs take up the driving thrust. Erik H. Delling, chief engineer of Mercer, had made exactly the same step in 1915. Since that day, and up to 1958 when coil springs and four-link suspension were made standard, Cadillac used Hotchkiss drive for all its V-8-powered cars. But when the V-12 and V-16 models were introduced in 1931, they had torque tube drive.

Studebaker went to Hotchkiss drive in 1920, due to Owen R. Skelton who had been a drive-train engineer at Packard for seven or eight years and later became one of the team members that created the first Chrysler. Chrysler products have always used Hotchkiss drive, and all its U.S.-made models except the Horizon still do.

Under the influence of Ormond E. Hunt, chief engineer of Chevrolet, and Henry M. Crane, technical advisor to GM president Alfred P. Sloan, all GM cars were gradually converted to Hotchkiss drive in the early 1920's-except for Buick, which retained its torque tube. Stutz made its last torque tube car in 1923 and went to Hotchkiss drive. Hudson adopted Hotchkiss drive for the 1924 Super Six, and Packard followed in 1927. Pierce-Arrow took the step from the torque tube to Hotchkiss drive in 1932.

It was the American example that drove European mass producers to switch to Hotchkiss drive, and Citroen, Fiat, Morris and Opel did it as if by general agreement in the late 1920's. They did it mainly because Hotchkiss suspension had the advantage of simplicity and low cost. Its geometrical and dynamic shortcomings were not significant in most cars that used it. Because they generally used very long leaf springs, which flattened out

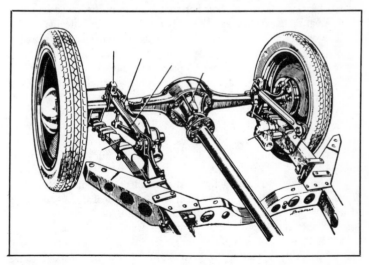

Fig. 15-2. In its original version, the Fiat 500 had quarter-elliptics and radius arms. This design was fine for road-holding with the driver alone, but not suited for heavier loads.

under load, the motions of the rear axle were mostly in harmony with both the propeller shaft and the front suspension (Fig. 15-2).

Cantilever and Platform Spring Digressions

Before rear suspension trends crystallized in this way, there were many sidelines that are worth considering. Two ideas that had an important following for some years were *Cantilever springs* and *Platform springs*.

A cantilever spring is a leaf spring with its center attached to the frame, carrying the axle at its extreme rear end. The front end is also anchored to the frame. The main advantage in this type of construction is that the spring itself is mainly sprung weight (instead of unsprung weight, as in Hotchkiss drive). Cantilever springs are capable of taking up the driving thrust, but they can also be used with torque tubes or radius rods.

First to use cantilever springs for a rear axle was the 1903 Lanchester, and the same basic design was continued through 1914. This was actually the first modern axle suspension, for the inventor, Frederick William Lanchester, a learned engineer, used a four-link parallelogram design with radius rods above the cantilever springs. This kept the axle housing in the same plane during spring deflections, avoiding the partial rotation that occurred with other systems. Similar layouts appeared on the 1911 Siddeley-Deasy and 1913 Lagonda. J. G. Parry Thomas made a more advanced version for the 1921 Leyland with double radius rods of equal length on each side and quarter-elliptic springs.

In a way, the 1901 curved dash Oldsmobile can be considered a precursor of cantilever springing for its body rested on two very long leaf springs forming a frame on their horizontal center sections and carrying the axles at their extremities. The Olds was a chain-drive design, while Lanchester used shaft drive. By 1906, Olds also went to shaft drive, using a heavy radius rod on the left side of the differential.

Few American auto makers adopted cantilever springs; but, in the 1920's, it became fashionable among builders of high-grade European cars such as the 1920 Napier 40/0, 1921 Hotchkiss AK, 1922 Farman, 1924 Lorraine-Dietrich, 1925 Rolls-Royce Phantom, and 1927 Renault 40 CV. Buick used cantilever springs in 1926, but soon reverted to its former arrangement. Another case was the 1916-21 Fergus which had cantilever springs with torque tube drive. King also used cantilever springs from 1921 to 24.

Many American constructors rushed to adopt platform springs. The principle involves a pair of longitudinal, semi-elliptic leaf springs anchored to the frame at their front ends, but shackled to the eyes of a transverse leaf spring at the rear. It was first used by Rolls-Royce in 1905 (on the 30 hp model) and then adopted for the Silver Ghost in 1906. Henry Royce did not depend on springs to take up the driving thrust, however, but added a long radius rod on the left side of the differential.

Platform springs gave extra freedom for the axle to move up and down, which was of great importance on the poor roads in the U.S. at the

224

time. Packard adopted platform springs in 1909; Peerless, Oldsmobile and Stevens-Duryea followed in 1910. It was a feature of the 1913 Heereshoff and a Cadillac as late as 1923. By that time, roads were getting better and simpler suspension could be used.

Rolls-Royce changed to torque tube drive with cantilever springs for the Silver Ghost in 1921 and adopted the same type of suspension for the 1925 Phantom. Then something strange happened. The company set up a branch factory in Springfield, Mass, and dispatched one of its leading engineers, Maurice Olley, to develop a U.S. version of the Phantom. He chose to use Hotchkiss drive, and it worked! By 1929 Henry Royce was eady to follow on his own disciple and fitted Hotchkiss drive on the Phantom II. All later Rolls-Royce cars up to the Silver Shadow (with independent rear suspension) have used Hotchkiss drive.

When Rolls-Royce took over Bentley in 1931, Royce's engineers discovered that Bentley had, in fact, been using Hotchkiss on all models since 1919, from the original 3-Liter up to the giant 8-Liter. This was surprising on the part of an old railroad-locomotive engineer, as W.O. Bentley was, and the explanation is that the rear suspension was not his idea, but the creation of his chief draftsman, F. T. Burgess, who had formerly worked for Huber.

Heavy Cars Went Torque Tube

On both sides of the Atlantic, most designers of heavy high-powered cars chose torque tube drive. All Voisin cars from 1920 to 1934 had a typical torque tube setup with semi-elliptic leaf springs. Maybach used splayed radius rods to help the springs locate the axle, while Hispano Suiza and Isotta-Franchini had designs very similar to Voisin's.

The Duesenberg A of 1920 had the same type of rear suspension system, which also resembled those of the Leyland-designed Lincoln. Lincoln retained the torque tube up to 1952. On the Marmon 32 and 234 the torque tube was combined with dual transverse leaf springs. Nash used torque-tube drive on most models from the start of 1917 right up to the merger with Hudson in 1954, and some AMC models retained this feature until a coil-spring and four-link system were adopted in 1967. Buick made a major change in 1937, going from leaf springs to coil springs and allying the torque tube with a system of diagonal struts plus a track rod for lateral location of the axle.

Many American cars had used 3½-elliptic springs to add some vertical wheel travel without the complication of platform springs (Rambler, Studebaker and Locomobile, to mention only a few of the most important). Some even used full-elliptics in the 1915-20 era, such as the Chevrolet Royal Main and Baby Grand.

Quarter-Elliptic Springs

Sports car engineers in Europe were exploring in the opposite direction. They reasoned that with cantilever springs, the front half of the spring is largely redundant, and thus came the quarter-elliptic leaf spring. Bugatti's Type 13 began its life in 1911 with semi-elliptics and a tapered radius rod from the axle casing to a frame cross-member. Two years later, the chassis was modified with reversed quarter-elliptic springs. The heavy ends were attached to the rear cross-member, and the front ends were tied to the axle. The driving thrust was thereby converted into a tractive force (Fig. 15-3). Bugatti adopted this suspension for the Brescia model and used it in modified form on a number of later models including the Type 35 single-seater, where the springs were supplemented by radius rods.

In 1914, Bugatti sold one of his designs to Peugeot, who then began using the same type of rear suspension on its light cars and continued that arrangement right up to the Peugeot 201 of 1929. The 1922 Austin Seven used quarter-elliptics located ahead of the axle, in conjunction with torque tube drive. Even better known for its quarter-elliptic rear springs was the Talbot. Made in London to designs by a Swiss engineer, Geroges Roesch, the Talbot matured from a small economy car to a powerful rally champion, and from 1924 to 1934 all models used rear quarter-elliptics -in the semi-cantilever configuration, not the reverse position of the Bugatti.

By the mid-1930's independent rear suspension systems, and the semi-independent de Dion type, had been brought to a high level of efficiency, and such designs offered valuable illustrations to engineers who were trying to perfect live axle suspensions. Fritz Fiedler of BMW invented a system of outstanding simplicity. A triangular bracket, attached to two points on the frame and to the top of the differential casing, took up the driving thrust. Together with the linkage to two longitudinal torsion bars, it secured the axle laterally and resisted torque reactions. But he did not use it on a pre-war BMW. Even the Type 328 had Hotchkiss drive. At the end of the war, Fiedler was invited to design a new car for Frazer-Nash, and here he put his torsion-bar design on a production model. The same arrangement was also fitted on the Bristol, and BMW began using its own version of the same design on its 1951 models (501 and 502).

The 1950 Jowett Jupiter, whose chassis was designed by Eberan von Eberhorst, had rear axle drive with a four-link system made up of upper and lower trailing arms. The upper ones were linked to transverse torsion bars.

Jaguar used Hotchkiss drive for the production-type XK-120 but adopted torsion bars and a separate torque-reaction member for the C-Type in 1951. The D-Type followed the same pattern, and then Jaguar went to independent suspension with the E-type in 1961.

Four-Link Radius Rods

The next dramatic improvement came from Ferrari, whose first touring car, Tipo 166 from 1947, had used Hotchkiss drive. The 1952-model 250 GT had an all-new chassis with two parallel radius rods per rear wheel, one above the other, and semi-elliptic leaf springs. The overall layout was basically the same as had been used on the 1921 Leyland. In

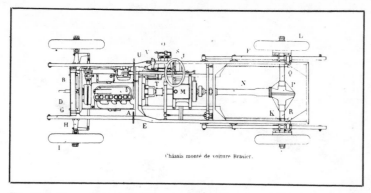

Fig. 15-3. Torque-tube installation on the Brasier circa 1911 shows how the driving thrust was transmitted to the frame cross-member aft of the gearbox.

1954, Ferrari went to a transverse leaf spring for the 500 Mondial sports car, retaining the same four-link radius rod design, and the following year the same layout was scaled-up to fit the new V-12 Ferrari 290 Mille Miglia.

Maserati and Aston Martin followed with intelligently worked-out rear axle suspensions later in the 1950's. The trend gradually spread into low-priced family cars, led by Volvo (Fig. 15-4), Fiat and Simca. Peugeot retained worm drive and a torque tube with coil springs through the 203, 403 and 404, changing to independent rear suspension for the 504 in 1969.

Ford, having gone to Hotchkiss drive across the board in 1949, went to a four-link rear suspension with coil springs in 1965. At the same time,

Fig. 15-4. Volva went to coil springs for their live axles on the PV-444 in 1944. This is the 144 version, from 1967, with trailing arms and coil springs.

227

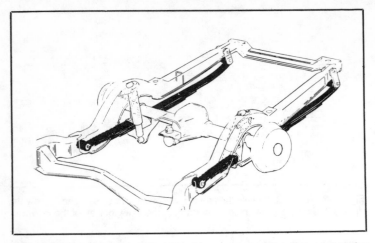

Fig. 15-5. Modern version of Hotchkiss drive, as used on Buick's largest station wagon in 1972 axle is clamped to forward portion of spring, which has an asymmetrical leaf arrangement.

Buick had developed a similar system to replace its torque tube drive (Fig. 15-5). The other GM makes went from Hotchkiss drive to coil-spring rear axle suspensions in 1958-59.

Chevrolet was first to use Hotchkiss drive with a single-leaf spring on the rear axle suspension for the 1962-model Chevy II followed by Ford's British-built Transit van in 1967.

Most high-grade European cars now have all-independent suspension. Will America follow? The 1959-64 Corvair and the 1960-63 Tempest failed because of flaws in the suspension geometry. The Corvette system works well, however, and could be adapted to mass-produced cars at modest cost.

Independent rear suspension with jointed drive shafts is only one of the alternatives to rear-axle drive. The other is front wheel drive. For the industry, it means a more costly initial change, but it may be a better long-term solution.

Chapter 16
The Story of Swing Axles

One thing that distinguishes the swing axle from other types of independent rear suspension is the constantly perpendicular angle of the drive shafts relative to each wheel, or vice versa. This feature alone puts the swing axle in a class by itself-fully independent in theory, but somehow non-independent in its jacking effect.

The swing axle seems historically identified with Porsche-from the Auto Union racing cars and the Volkswagen to the first cars bearing the Porsche name-but the swing axle was not, in fact, a Porsche invention.

The swing axle was invented at the beginning of the century be a famous auto and aircraft engineer, Edmund Rumpler, of Austrian birth but who spent practically his entire working life in Germany. Why he thought of the swing axle so early in the evolution of the automobile is hard to say. With chain drive, which was then predominant, a live rear axle avoided torque reactions in the lateral plane, so there was less incentive to try for independent rear suspension than with shaft drive. Knowledge of Louis Renault's shaft drive spread so quickly throughout Europe (partly due to the Renault brothers driving their own cars in major international races) that it is inconceivable that Rumpler did not know about it.

Rumpler's Inventive Approach

Ride comfort and roadholding ability were hardly thought of as consciously sought-out goals in those days. It is doubtful that Rumpler was even concerned about unsprung weight. On the other hand, there can be no argument that reliability was the main objective; and, in most cases, engineers tended to equate reliability with simplicity. Cost was not a major consideration. But simplicity of design meant not just lower cost; it meant that manufacturing and assembly were made simpler, with less to go wrong.

Let us assume, therefore, that Rumpler had examined Renault's shaft-driven axle and thought it too complicated. Perhaps he saw the drive-shaft pinion as a natural center, a meeting point of symmetrical units using identical components. From there, each half-axle could swing in a pendulum pattern on wheel deflection without causing stress peaks in the drive train and without causing wear and play in gears, joints (of which there were none) or suspension members.

Rumpler had the idea of taking the final drive unit out of the rear axle and mounting it in the chassis frame, so that the drive line between the engine gearbox and differential would be unaffected by suspension movements. Rather than using universal joints, either inside the differential or outside, Rumpler chose to use a separate ring gear for each rear wheel. But they could not be driven from the same pinion, for they would then revolve in opposite directions. This problem was solved by the clever method of extending the propeller shaft beyond the first pinion and fitting a second one that meshed with the second ring gear. Clockwise rotation of both pinions then resulted in forward rotation of the first ring gear (meshing left of center) and forward rotation also in the second (meshing right of center).

Each ring gear was mounted at the inner end of a drive shaft that was enclosed in a casing that was integral with the wheel hub housing. At the differential end, each shaft carried a segment inner sleeve that permitted partial rotation within the final drive housing so as to allow for wheel travel. Tilting of the axle shafts made no difference to the meshing between the pinions and ring gears, it just displaced the contact surfaces fractionally ahead or backwards in time, according to whether the travel was in jounce or rebound.

Near the wheel hubs, longitudinal semi-elliptic springs were attached, assuring fore-and-aft wheel location (as in Hotchkiss drive). The use of dual ring gears gave maximum length to the drive shafts so as to minimize camber changes, and the spring arrangement provided pure lateral geometry, in the sense that wheel travel did not result in either toe-in or toe-out changes.

Rumpler Patent Adaptations

Rumpler was hired by Adler in Frankfurt in 1902 to design a new range of cars, and it was in 1903 that he filed his patent for the swing axle. The patent was assigned to Adler, but it is uncertain whether Adler produced any swing-axle models for sale. Rumpler remained with Adler till 1906 when he moved to Berlin to work as an engine consultant and experimenter with aircraft, temporarily dropping out of the automobile scene.

Ferdinand Porsche, an intrepid driver as well as a brilliant engineer, began to experiment with swing axles in 1909 when he was chief engineer of Austro-Daimler. This work had no dire ct result on the production models, but impressed a young engineer, Karl Rabe, so favorably that he,

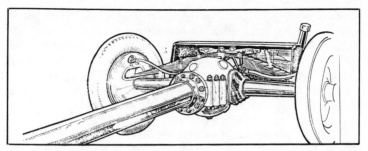

Fig. 16-1. High-powered Austro-Daimler from 1927 also took up the driving thrust via the inboard axle bearings. Transverse leaf spring tended to keep wheels always in positive camber position.

having succeeded Porsche in 1923, would use swing axles on the Austro-Daimler ADR in 1928 (Fig. 16-1). By that time, however, he had much more study material at hand-notably in the works of Max Wagner at Benz and Hans Ledwinka at Tatra.

After World War I, Rumpler returned to car design and built a number of radical machines with inboard rear engines and swing axles. In 1922, Rumpler sold Benz a license to his swing axle and chassis design, and it was used on the 1923 Tropfenwagen in combination with splayed cantilever springs whose rear ends were connected to the swing axle casings near the wheel hubs. The Tropfenwagen was a streamlined single-seater, intended for Grand Prix racing, of which only a half-dozen were built before the project was scrapped in 1926.

Though made under Rumpler patents, Max Wagner discarded the dual ring gear in favor of universal joints on each drive shaft close to the differential. The shafts were enclosed in casings resembling torque tubes, with pivot points ahead of and behind the universal joints. The wheelhub housing was, as in the Rumpler patent, integral with the casing. The splaying of the springs gave a slight toe-in on both jounce and rebound which was probably intended to counteract the inherent oversteering tendency in the car. Also, Wagner's use of universal joints shortened the swing radius of the wheel and axle shaft and aggravated the camber changes. But the springs were stiff (as we thought necessary for a racing car at the time) and wheel travel was short.

Ferdinand Porsche was now at the Daimler Motor Works as chief engineer for Mercedes cars. There was not yet a connection between Benz and Mercedes, though the companies formed an interest union in 1924 that led to a full merger in 1926. As early as 1924, however, Porsche was again experimenting with swing axles, using some of Wagner's techniques on Mercedes test cars.

Ledwinka, Tatra and Steyr

Hans Ledwinka chose the Rumpler-type swing axle for his Tatra T-11 and T12 of 1923 (Fig. 16-2), using dual ring gears in the final drive; but he went to a wide-transverse leaf spring, located above the axle shafts. These cars were built up on a central steel-tube frame with the propeller shaft inside it. The driving thrust was taken up by the bearings that held the half-axle casings at the inner joints—tolerable for low-powered cars but frivolous for high-performance machinery. The Tatra T-11/12 engine was a flat-twin located up front, and the cars had I-beam front axles. The same layout was adopted for the T-30 in 1930, the small T-57 in 1931, plus a series of Tatra trucks beginning in 1925.

Ledainka was an uncommonly prolific engineer who had been responsible for the first Steyr car of 1918 which had a conventional chassis with a live rear axle and cantilever springs. When the management of Steyr saw the new Tatra models, the Austrians wanted swing axles for their next car and made a deal with Tatra and Ledwinka for the design. Steyr had a conventional frame and would not accept the backbone frame of the Tatra. Ledwinka came up with an admirable solution, moving the universal joints into the center of the differential and using a single ring gear. The wheel hubs were located by trailing radius arms that extended behind the hubs and provided attachment brackets for the ends of a transverse leaf spring. This setup was introduced on the Steyr models up to about 1935.

Porsche, Rohr and Nibel

In 1929, Steyr had a new technical director-none other than Ferdinand Porsche, who had left Mercedes. He inherited a line of Ledwinka-designed cars and proceeded to change them first, intending to replace them gradually with his own at later dates. Porsche's first swing-axle design for Steyr was used on the XXX of 1930 and used a high-mounted transverse leaf spring. Then Steyr merged with Austro-Daimler,and Porsche went back to Stuttgart to set up a private engineering office.

Karl Jentschke emerged as Steyr's new engineering chief after that, and he developed a series of cars, starting with the 125 and including the

Fig. 16-2. Swing axles with twin ring gears (for single central pivot point) on 1923-model Tatra 12 put driving thrust loads into differential bearings.

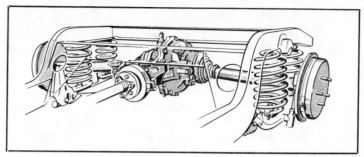

Fig. 16-3. Split pivots and dual coil springs were used with the swing axles on the 1931 Mercedes-Benz 170. Later models had coil springs within the coils, one set on each side of the wheel hub.

200 and 220, that retained the swing axles but combined them with longitudinal quarter-elliptics doing double duty as radius arms.

Hans Gustav Rohr, after failing to get established car makers to buy his designs, finally made up his mind to become a manufacturer himself and got production of the Rohr 8 under way in 1927. This car had a swing-axle rear end with universal joints on either side of the final drive and cantilevered semi-elliptic leaf springs. It is believed that Rohr got the idea from Rumpler. The detail design was due to Rohr's assistant, Joseph Dauben, who must get credit for the quality and handling characteristics of the car.

Among inquisitive rivals who scrutinized the Rohr chassis were both Ferdinand Porsche and Hans Nibel. Nibel had joined Benz in 1904 and became its chief engineer in 1908. Max Wagner, who joined Benz in 1910, was in charge of special projects, from diesel-engine research to racing cars, and Nibel had watched Wagner's experiments on the Rumpler patents with considerable interest.

Nibel began to evolve his own theories on independent front and rear suspension. He began to sketch swing-axle systems with various types of springs and geometry. What this work synthesized into was a system with torque-tube-type swing axles, oscillating up and down on trunnion-bearings which transmitted the driving thrust to the differential casing (which, in turn, was mounted in rubber blocks in the frame). For the elastic element, Nibel chose dual coil springs, mounted vertically-one coil ahead of and one coil behind each rear wheel hub. This design was first used on the 170 of 1931. It was adapted for the 200 and 290 of 1932, 380 of 1933, 500 of 1934 and 540 of 1936.

The same basic layout was chosen for the W-25A Grand Prix car of 1934-35, where the coil springs were replaced by transverse quarter-elliptics. Nibel died in 1934 and chassis developmet at Mercedes-Benz came into the hands of Max Wagner. Wagner went to de Dion suspension for the W-125 in 1937 and the Type 700 (Grosser Mercedes) in 1938, while retaining swing axles and coil springs for the 230, 320 and—with single coils—the new four-cylinder 170V (Fig. 16-3).

Porsche's VW Ancestor

The second swing axle from Ferdinand Porsche's drawing board appeared in a small-car project for Zundapp in 1932, and the same design was part of a similar car he did for NSU in 1933. These forerunners of the Volkswagen had half-axles with joints outside the differential and torsion bars. The torsion bars were positioned longitudinally and operated by lateral rods. Porsche executed a similar swing-axle layout for a medium-size, front-engined car with a conventional frame, which he sold to Mathis. It became the Emysix of 1933 (which also had Porsche's independent front suspension with trailing links and torsion bars).

When Porsche designed the first Auto Union racing car in 1933, he chose a transverse leaf spring mounted above the final drive and swing axles, with diagonal radius arms pivoted in brackets on the narrow-base tubular frame. The geometry of this layout was akin to that used on the Benz Tropfenwagen, in fact the swing axles did not move up and down in a purely vertical plane, but pivoted on an axis 23 degrees from longitudinal (made up by drawing the line between the universal joint center and the radius-arm pivot point).

For the 1935 Auto Union, Porsche went to torsion bars and a 35-degree pivot angle, and this basic setup lasted through the C-type of 1936-37 (Fig. 16-4). Then Auto Union severed its connection with Porsche, and William Werner, who took responsibility for chassis design, went to a de Dion axle for the 1938 3-liter supercharged V-12.

Skoda, Hansa and Borgward

Skoda and Hansa took up using swing axles on light, four-cylinder family cars in 1935 and 1936, respectively, using transverse leaf springs and trunnion-bearings around the axle joints to take up the driving thrust.

Fig. 17-4. Swing axles on the V16 Auto Union of 1936 were linked to longitudinal torsion bars and had diagonal radius arms for taking up the driving torque.

The first Skoda so-equipped was the Rapid, which lived on to 1939. It was revived in 1946 as the Skoda 1100, becoming the Octavia in 1953. When switching to rear-engine mounting for the 1000 MB, Skoda added radius rods for fore-and-aft location of the wheels. That design from the late Fifties is still in use at Skoda.

Hansa used swing axles on the 100 and 1700, which were built up to 1940. The Hansa owner, Carl F. W. Borgward, came back in 1954 (Fig. 16-5) with a new car, the Borgward Isabella, using the same type of swing axle. The Isabella lived on until Borgward went under in 1961.

Alfa Romeo and Ferrar

Alfa Romeo used swing axles on the 8C of 1935 and 12 C of 1936, both single-seater racing cars with diagonal radius arms and transverse leaf springs. Alfa Romeo had an arrangement with Porsche, who was not only an engineering consultant, but also furnished new designs to the nationalized Milan firm. Also in 1935, Alfa Romeo put swing axles on a production car, the 6 C 2300 B. This design also used a transverse leaf spring and was carried over into the 1947 6C 2500. The Type 158 Alfetta Grand Prix car of 1938⅜9 had swing axles and was raced from 1946 to 49 with great success. It was redesigned in 1950 with a de Dion rear end and lap speeds became even faster. That ended Alfa Romeo's involvement with the swing axle; most subsequent production models, beginning with the 1900 in 1950, have had live rear axles.

Ferrari, familiar with Alfa Romeo practice for decades, chose swing axles with a transverse leaf spring for his first Grand Prix car, Type 125, of 1947, no doubt feeling that the Hotchkiss drive of the Type 166 sports car would not be sufficient. But when Ferrari dropped the supercharged 1,500-cc V-12 in favor of the unblown 4-1/2-liter in 1905, the swing axles were discarded in favor of a de Dion axle.

First VW and Porsche, then Renaul

The original Volkswagen, a Porsche design identified as starting with the Type 60 of 1934, had swing axles from the earliest design stage. It borrowed the diagonal pivot axis from more powerful cars in combination with longitudinal torsion bars. By 1935, however, the torsion bars had been relocated transversely, and flexible trailing arms had taken the place of the radius rods. The Pivot axis went from a 19- to a 40-degree angle. This setup was used on the final VW Prototypes in 1939 and on the first production cars in 1945. The same layout was also adopted for the first Porsche, made in Austria in 1948 mainly from VW components, with its own sports-car body. Porsche went to trailing-arm suspension (rear) for the 911 in 1967, and the Volkswagen Beetle was redesigned with semi-trailing arms for 1969.

When Louis Renault saw the Volkswagen prototype in Berlin in 1939, he and his chief engineer, Edmond Serre, started to copy it. The result was

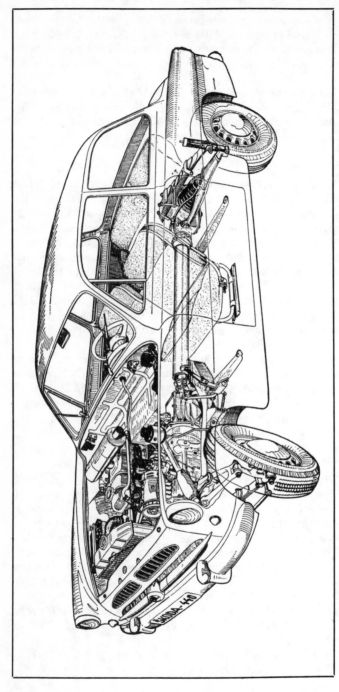

Fig. 16-5. Skoda Octavia of 1954 used swing-axle casing with widely splayed supports at the inboard end, front and rear of the universal joint and transverse leaf spring.

the 4 CV which came out in 1947. It had swing axles, but with three notable differences: 1) Renault's car reverted to trunnion-bearings for taking the driving thust, eliminating the trailing arms, 2) it had much smaller wheels than the VW, which augmented the camber changes on deflection, and 3) Renault used coil springs instead of torsion bars. This design lived on in the Dauphine (1956), R-8 (1962) and R-10 (1965). On the 1962 Floride (exported with a Caravelle label) Renault did add radius arms (Fig. 16-6). By that time, Renault was well into its full transition to front wheel drive, and the swing axles disappeared about 1972.

Long Adoption by Mercedes

Mercedes-Benz got back into civilian production in 1946, continuing the pre-war 170 V. It was redesigned for 1949 as the 170S. On the first all-new, post war design-the 180 of 1953-chief engineer Fritz Nallinger added radius rods. This model also introduced the vertical, single-point locating tower for the final drive casing, with a high-volume rubber bushing at its top (body) end.

Alfa Romeo had been the first to demonstrate that it was almost a necessity to keep the wheels constantly on the negative-camber side in order to maintain cornering stability and handling precision with swing-axle-equipped cars. Following the Italian example, Nallingen and his chief development engineer, Rudolf Uhlenhaut, began to cut back on positive

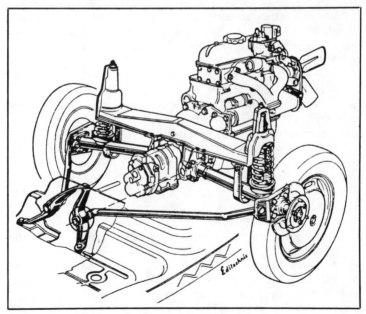

Fig. 16-6. Renault R-8 of 1962 added diagonal radius arms to the swing-axle rear suspension inherited from the 4 CV and Dauphine.

camber travel and letting the wheels run deeper into the negative field. But staying clear of positive camber put a severe limit on wheel travel, and they were forced to add auxiliary spring systems to keep from bottoming or excessive negative camber on the wheels, raising the spring rate and preserving a certain jounce travel margin. Tpe torsion bars had splined arms on their front ends, and the electric motor was gerred to a rack that pushed out stops to contact the arms when needed and retracted them automatically when the overload was removed.

The 200 SL 'Gullwing' coupe of 1954 began life with the same arrangement, but the 300 SL convertible of 1957 had a transverse horizontal coil spring above the final drive unit. This later model was also equipped with the low-Pivot-point swing axle which had been introduced on the 190 SL of 1955. It had originated on the 1954 W-196 with open drive shafts and a Watt linkage for fore-and-aft location of the wheels.

In this design, which is usually credited to Hans Scherenberg, the swing-axle function was divorced from the half-shafts. The hub carriers were formed as uprights, bolted to curved, lateral swing arms sharing the same pivot shaft on the car's longitudinal center line. These arms took care of the lateral location of the wheels. Their great length helped minimize camber changes on deflection, and the roll center was located at their pivot axis, below the differential.

Bell-cranks attached to the lateral swing arms were coupled to torsion bars running forward to anchorage points in the cowl structure , and telescopic shock absorbers were mounted vertically on brackets extending backwards from the swing arms. The Watt links were equal in length, one running forwards from a joint at the bottom of the hub carrier, and the other backwards from a joint at the top.

It became the task of Scherenberg and Uhlenhaut together to redesign and develop this low-pivot-point geometry for production cars. Instead of four universal joints, the final design had only one-on the center line. The differential and its casing were part of the left swing axle, undergoing the same oscillations. A rigid arm extended downwards, front and rear, from the differential casing to a pivot shaft whose center section provided bearings for the forked end of the right-hand side-axle casing. This fork was provided with roller splines to permit length variations in the shaft. The pivot shaft also had bearings for a tower bracket with a fat rubber bushing on top, at its anchorage to the body shell.

For the 190 SL, vertical coil springs were mounted on radius arms, without a camber-compensating spring. This was added for the 300 SL and also adopted for the 1959-model 220. It was used on the 230 SL in 1963 and its successors, plus the W-108 (250 sedan) of 1965. But with the next generation, the W-114 (220, 230, 250 and 280) of 1967, Mercedes-Benz went to double-jointed drive shafts and semi-trailing arms, burying its swing-axle designs in the archives.

Incidentally, the Porsche engineers started to get worried by excessive positive camber about 1955. The result was a transverse single-leaf spring, conically rolled towards the ends, which were linked to the hub carriers. At the center, it was flexibility attached to the differential casing. It was first used on the 356 A Super 90 about 1957 and remained in use for all subsequent 356's.

Kieft and Fiat Ideas

The single-pivot idea, which Rumpler had at the outset, was taken up by Ray Martin and John A. Cooper, co-designers in 1950 of the Formula 3 Kieft for Cqril Kieft. The swing axles were asymmetrical with the pivot point slightly offset to the right, though not lowered relative to the differential center. However, as the rear wheels on this car were set to run with pronounced negative camber, the pivot point was effectively lowered in relation to the wheel-hub centers. Gordon Bedson, who joined Kieft in 1953, took this swing-axle idea with him to Raymond Flower's enterprise in 1955, where he designed the Frisky minicar.

Under the direction of Oscar Montabone, Fiat had been experimenting with novel types of swing axles for minicars, and about 1950 a design evolved which separated th e drive shafts from the swing arms, as on the Mercedes-Benz W-196. Fiat's design was cruder and cheaper, since it was intended for a low-priced,mass-production car.

The hubs were located by a combination of lateral swing arms and longitudinal radius arms. The drive shafts and universal joints were out in the open. This design was first used on the Fiat 600 of 1955, with a wide pivot-axis angle, 37 degrees from longitudinal, giving strong toe-in in jounce, with negative camber. It was also adapted for the even smaller Nuova 500 of 1957, with the pivot-axis angle increased to 43 degrees. But with the arrival of the 850 in 1964, Fiat went to double-jointed drive shafts and semi-trailing arms.

Variety of Adopters

A new model of the Isuzu Bellett, appearing in 1964, used a Fiat-type swing axle design. But Isuzu went back to a live rear axle by 1969.

The first DAF cars from Eindhoven used swing axles in combination with Van Doorne's Variomatic belt drive. The first 600 of 1958 had a coil spring system which was carried over unchanged for the 750 and improved for the Type 44 in 1963. But with the introduction of Type 55 in 1968, DAF went to semi-trailing arms.

Another interesting variation on the swing-axle theme had appeared in 1955 using open, unsplined drive shafts that worked as lateral locating members, transmitting lateral stresses through the universal joints. Fore-and-aft location was assured by radius rods, and the elastic element was a pair of coil springs. This design by Karl Dompert appeared on the 1955 Goggomobil (later Glas) which lived on to 1966.

This design may have inspired Harry Webster at Triumph to use open, unsplined, stress-carrying shafts in the swing axle suspension for the Herald in 1958. The hub carriers were formed with an upward extension leaning inward to provide attachment points for the ends of a high-mounted transverse leaf spring, with telescopic dampers tilted about 30 degrees from vertical behind the hub. On the front side, radius arms pointed slightly inboard, mounted on rubber bushings at both ends.

The same setup was used for the Triumph Vitesse of 1962, Spitfire of 1963 and GT-6 of 1966. Three years later the GT-6 and Vitesse rear ends were modified by the insertion of a Rotaflex rubber doughnut joint in the drive shafts at the wheel hub end, and the addition of a lower lateral arm. But the Spitfire continued with swing axles and still uses the original layout.

Corvair Led Limited U.S. Use

America's auto industrq, always attached to the live rear axle, never showed much interest in any form of independent rear suspension for the longest time. But when Ed Cole at Chevrolet chose to put the engine in the tail end of the first compact, the Corvair could not possibly accommodate a live axle. A former Mercedes-Benz engineer named Robert Schilling, who joined General Motors, was in charge of designing the chassis for the 1960 Corvair and created a swing-axle system with geometry resembling that of the Fiat 600. The Corvair had semi-trailing arms and coil springs. The

Fig. 16-7. Pontiac Tempest swing axles had semi-trailing arms for handling the driving thrust. The roll center was 14.5 inches above ground level.

swing axle design used on the experimental Firebird II of 1956 (gas-turbine car) was not Schilling's, but Joe Bidwell's, and probably did not contribute anything to the Corvair development.

The Corvair swing axles were adapted for the 1961 Pontiac Tempest (Fig. 16-7) which had a front engine and could well have used the same live axles as the Oldsmobile and Buick compacts. But the man who was head of advanced engineering at Pontiac at the time, John DeLorean—a very bright guy, was dominated by an innovative spirit. He wanted the swing axles, but also created the curved drive line and put the transmission in unit with the differential rather than with the engine.

But the Tempest and Le Mans lost their swing axles in 1964 when they became intermediates, rather than compacts, going to live axles. And for 1965, the Corvair got a new rear end with double-jointed shafts, a far superior arrangement which lasted until the Corvair's demise in 1969.

Chapter 17
Modern Independent Rear Ends

Three families of independent rear suspension design fit the requirements of what we now call modern. One family is derived from front suspension design and uses lateral A-arms in one way or another; the second family can be described as an adaptation of MacPherson front suspension principles to driven rear wheels; and the third family originated with driven wheels and uses trailing arms or semi-trailing arms.

Chronologically, the first design of the lateral A-arm family appeared in 1922. It was designed by Maurice Sizaire and was basically a duplication of the independent front suspension system of the Sizaire Freres car, with an upper A-arm and lower transverse leaf spring. The final drive housing was bolted to frame cross-members, and the drive shafts had inner and outer joints.

A contemporary Italian car, the little San Giusto, had equal-length upper and lower A-arms and a transverse leaf spring on top of the whole suspension assembly (Fig. 17-1). This car had a central-backbone frame which carried the midships engine and final drive as a unit, with double-jointed drive shafts to each rear wheel.

For the next example, we return to France. The 1927 Cottin-Desgouttes carried the rear-wheel hubs on four transverse leaf springs-an adaptation of the front-wheel drive Alvis principles to driven rear wheels (Fig. 17-2) .

In connection with its major swing to independent front suspension, General Motors in 1933 was making serious studies of independent rear suspension also. Many leading design and experimental engineers-including Maurice Olley, Alex Taub, and Olle Schjolin-were involved with this research program, but it is not known today who did exactly what. A

Fig. 17-1. Rear suspension of the 1923 San Guisto was copied on its front end, with the addition of double-jointed drive shafts.

prototype car known as the Albanita had, at one stage in its life, a rear end with upper and lower A-arrs of equal length , with pivot shafts anchored in a cage carried at the end of the backbone frame and enclosing the final drive unit. Dual coil springs were used at each rear wheel. Test work with the Albanita went on, at least, up to 1938.

Few Lateral A-Arms Until Sixties

The 1935 R-Type MG designed by H. N. Charles could almost be described as a front-engined San Giusto with a rear suspension made up of

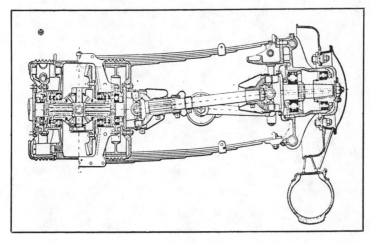

Fig. 17-2. Quadruple-leaf springs made a square-section cage around drive shafts at the rear end of the 1927 Cottin-Desgouttes. Drive shafts were double-jointed, and springs acted as equal-length lateral A-arms.

243

equal-length A-arms, the main difference being the use of torsion bars instead of the transverse leaf spring (Fig. 17-3). The 1936 front-wheel drive Lloyd had a similar rear suspension.

John N. Cooper, working on a 500 cc racing car in 1945, copied the Fiat 500 front suspension, with a lower A-arm and upper transverse leaf spring, for both front and rear ends (Fig. 17-4). John Tojeiro adopted the same idea on a larger scale, since the AC was powered by a two-liter six) when he designed the AC Ace in 1952.

Cooper later switched to coil springs, setting a trend in racing car engineering that continues to this day. Lotus, BRM, Maserati and Ferrari followed. In 1959, even Porsche abandoned the swing axle on its Formula Two car and adopted lateral A-arms with torsion bars).

In 1960, Colin Chapman introduced a new idea on the Formula Two Lotus: that of using the double-jointed drive shaft as the upper stress-carrying location member. Actually the idea was not new, for it had been patented by Georges Roesch of Talbot in 1933; but it had not been used until Chapman re-invented it. The Lotus had lower A-arms with diagonal radius rods at the upper and lower ends of the hub carriers, and inclined coil springs leaning inboard from the joint linking the A-arm to the hub carrier to its upper mounting on the tubular chassis frame.

Fig. 17-3. MG R-Type from 1935 had independent rear suspension with equal length A-arms and longitudinal torsion bars. Brake drums were mounted outboard.

Fig. 17-4. Rear suspension of the 1952 AC Ace evolved from the Fiat 500 front end of 1935 via the 1946 Cooper F-3, which had a similar setup at both ends. Cooper and AC dispensed with the upper control arm, letting the transverse leaf spring serve as a locating member.

Major Jaguar and Corvette Designs

William Heynes of Jaguar borrowed this idea for the E-Type of 1961, and Chevrolet's Zora Arkus-Duntov used it for the 1963 Corvette. The E-Type was the first production Jaguar to use independent rear suspension. Early design and development work on this innovation included a small military vehicle with swing axle, independent rear suspension and a rebuilt D-Type with a de Dion rear end. Experience with both led inescapably to a new design with lateral control arms.

The XK-E design included a massive, lower triangular arm, while the drive shaft acted as an upper arm in the Lotus manner. The lower arm was mounted in widely-spaced roller bearings on the lower part of the rear suspension sub-assembly which also carried the final-drive unit. The outer end of the lower arv was pivoted on a pair of tapered roller bearings, also widely-spaced and carried at the lower end of the cast-aluminum hub carrier. The hub carrier was also attached to a radius rod which took up the driving thrust. The radius-arm pivot points had rubber bushings, providing a small amount of horizontal compliance.

The same setup was used for the Mark X limousine in 1962 and for the S-Type sports sedan in 1963. An entirely new design with dual vertical coil springs and wider A-arm was prepared for the XJ series of 1968 and later adopted for the XJ-S.

The Corvette rear suspension system has a lot in common with Jaguar's. Both use unsplined axle shafts as upper control arms and have simple transverse lower-control arms with longitudinal forces taken up by parallel radius arms. But while Jaguar uses coil springs, the Corvette has a transverse leaf spring. The decision to use a leaf spring was made mainly for reasons of space availability-a factor which also limited wheel travel. The original design called for 3.75 inches of wheel travel in jounce, but the fender line dictated by GM Styling caused a reduction to 3.15 inches.

The spring mass is carried as sprung weight on the chassis (the flexing leaves are semi-sprung) and the change from a rigid axle to fully-independent suspension in the rear end of the Corvette brought about an improvement in the sprung-to-unsprung weight ratio from 5.27:1 to a very favorable 7.98:1. The nine-leaf spring uses full-length polyethylene liners to provide constant interleaf friction. The spring ends are linked to the hub carriers by rubber-cushioned vertical link bolts, and hub carriers are integral with the massive radius arms. This system proved extremely successful both in general service and in competition, and for 1965 the same geometry was applied to the Corvair chassis in combination with coil spings.

MacPherson and Other Spring Legs

The 1960 F-2 Lotus did not have what has later become known as a Chapman strut, which had made its first appearance on the Lotus Elite in 1958. Actually it is a misnomer, for the Chapman strut is nothing but a MacPherson spring leg. The original MacPherson patent drawings show its use in the suspension system of driven rear wheels. This was the design intended for the stillborn Chevrolet Cadet, shelved in 1947,

Here, the strut does not have to contain rotating parts, since rear wheels do not steer. Its base can be fixed to the hub carrier and its upper end anchored in the body strucrure, The drive shafts are double-jointed, but completely relieved of locating duties. Variations in length were accommodated by Metalastik rubber doughnuts serving as inner univerapsal joints. In the original Elite design, a cranked radius rod ran diagonally forward from the hub carrier to assist in handling the driving thrust. But in 1960 the cranked radius rod gave way to a semi-trailing arm-not so much for reasons of geometrical design as for durability considerations. The loads put into the radius rod tended to cause damage to its mountings in the fiberglass body structure. Chapman used the same strut suspension on the Lotus Elan in 1963, and Alfa Romeo adopted it for the Giulia GTZ in 1964.

While the MacPherson spring leg enclosed a telescopic shock absorber and carried a concentric coil spring on its upper portion, other types of spring legs have also come into use providing the same geometrical advantages.

In 1969, Fiat introduced its 130 with vertical guide struts and diagonal radius arms, using the double-jointed drive shafts as lower control arms. The guide struts were combined with the shock absorbers, but the weight was carried by large coil springs mounted on the radius arms. The same year Fiat also introduced the 128, which had front-wheel drive and rear wheels supported by a transverse leaf spring and strut guidance instead of an upper control arm. This design was adopted in 1978 for the RitmoT-RADA. Abarth developed an independent rear end for the Fiat 124 Spider

246

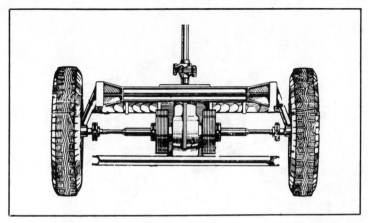

Fig. 17-5. Trailing arms on the Lancia Aprilia of 1937 were linked to transverse torsion bars plus a transverse leaf spring. Brake drums were positioned inboard to reduce unsprung weight.

in 1974, using Chapman struts, and this setup was adapted for the rally version of the 131 Mirafiori in 1976.

Lancia First with IRS Trailing Arms

The idea of trailing arms is, of course, akin to Porsche's invention of the trailing links. One noteworthy difference is that trailing links form a pair, of equal length, which acts, as a parallelogram; a trailing arm is a single component with a pivot axis formed by a shaft of a certain length, or separate pivot points spaced at a certain distance. While the pivot points of trailing links are placed at different levels, the trailing arm pivot axis is horizontal or near-horizontal.

In 1936, Cord had trailing arms for the driving wheels of the 810, which were at the front, but the same basic design could have served well for driven rear wheels also. Citroen has always used trailing arms for the rear suspension of its front wheel drive cars, from 1934 to date.

It was Vincenzo Lancia who created the fist independent rear suspension system for driven wbeels with trailing arms on the 1937 Aprilia (Fig. 17-5). A transverse leaf spring located near the pivot axis of the trailing arms supported the static weight, while transverse torsion bars extending from the pivots were intended to resist bump travel and body roll. The trailing arms were connected to the leaf spring ends by flexible cables. Drive shafts were double-jointed and splined.

For the Aurelia, which replaced the Aprilia in 1950, Lanci's chief engineer, Vittorio Jano, adopted semi-trailing arms, fabricated as triangles from steel tubes, and discarded the transverse leaf spring in favor of coil springs. The pivot axes for the two anchorage points were not in alignment, and that was for a purpose (Fig. 17-6 and 17-7). The swivel points were mounted in rubber bushings that were compressed whenever

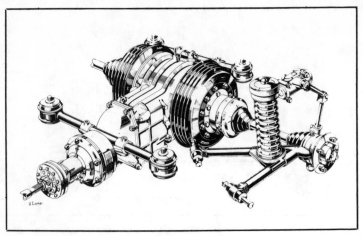

Fig. 17-6. The first semi-trailing-arm design was used by the Lancia Aurelia of 1951. The transaxle had four-point resilient mounting in the body structure, and brake drums were mounted inboard.

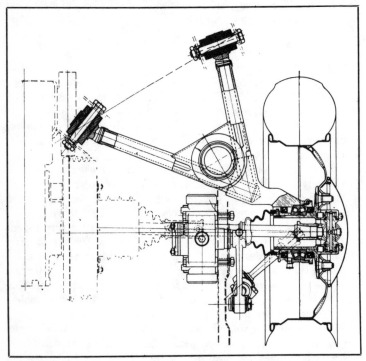

Fig. 17-7. Pivot points of Aurelia's semi-trailing arm were not aligned and worked only by dint of flex in the rubber bushings. True pivot angle was 60 degrees from transverse axis.

the wheel was deflected from its normal (design-height) position. This provided a measure of progressivity in the resistance to wheel travel, up and down. In fact, if it weren't for the flexibility of the rubber bushings, the system would not have worked. The semi-trailing angle of the arms gave toe-in in both jounce and rebound, with increasing negative camber in jounce. (Fig. 17-8)

Drive shafts were two-jointed to give the wheel freedom to follow the camber angle dictated by the semi-trailing arms. Coil springs were mounted vertically as close to the wheel hubs as possible, with piston-type

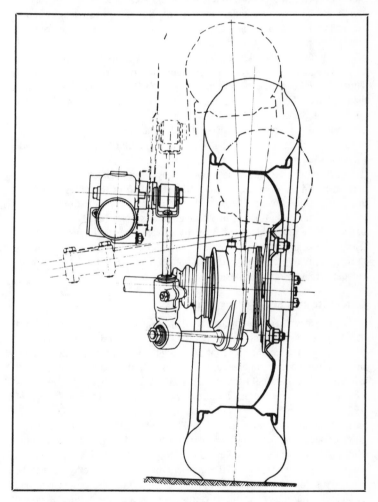

Fig. 17-8. Rear wheels on the Aurelia went into negative camber angles in jounce, which also increased toe-in. Shock absorber linkage had freedom to work in three planes without interference.

hydraulic shock absorbers linked to the hub behind the drive shafts. This design gave way to a de Dion rear end in 1953.

One memorable design that really belongs in the semi-trailing arm family no doubt started out as a type of swing axle. This was the 1947 Lagonda, designed by W. O. Bentley. The rear wheels were driven by double-jointed, splined shafts; the hub carrier held two flanges, one serving as the base for a conical tube whose other end was pivoted in a ball-joint behind the differential, the other providing rear anchorage for a diagonal radius arm running forward to a ball-joint on the frame. These members combined to form a triangular control arm. Because of the flanges being integral with the hub, camber and toe-in changes were fixed by the control arm position, and because the members were long, these changes were relatively slight. The hub was also linked to a short lateral arm which twisted a diagonal torsion bar located inside the cruciform frame structure. There was no second control arm. Since the ball-joints were located on a diagonal axis, the Lagonda rear suspension can, in fact, be categorized as a semi-trailing-arm system.

Another car which appeared at the same time, but never got into production, the Invicta Black Prince, had something approaching Mac-Pherson geometry in the rear end, though the components were totally different. The wheel hub was supported by a big horizontal bracket that carried a sliding pillar on each side (front and aft of the outer universal joint). The pillars were pivoted at the top end. Below the outer universal joint, the hub was attached to a triangular control arm, short but wide-based, whose pivot axis was formed by a torsion bar extending forwards. The drive shafts were two-jointed and splined, relieved of all locating duties. No doubt this design by William G. Watson worked well; but, in retrospect, it seems needlessly complicated.

Current Trend to Semi-Trailing Arms

The present trend is towards semi-trailing arms-a trend that was born at BMW, where the first of the modern designs matured on Fritz Fiedler's drawing board in 1961. Apart from studying the Lancia Aprilia and Aurelia, he may have looked at-and rejected-the narrow trailing arms and horizontal coil springs of the 1939 Atalanta, as well as the wide-based trailing arms with vertical coils designed by Fernard Picard for the Renault Fregate of 1951.

The rear suspension of the 1963-63 BMW 1500 consisted of two pressed-steel semi-trailing arms with vertical coil springs and telescopic shock absorbers mounted behind the springs and tilted inboard. The arms were mounted at an angle of 20 degrees to the transverse center line of the car. At a side force of 0.5 g, both front and rear outside wheels assumed about 1.5 degrees of positive camber, with roughly similar negative camber angles on the inside wheels. The roll center was located 5.15 inches above ground level.

No major changes have been made in this basic design, which has been adapted for all BMW cars produced during the years Bernhard Osswald ran the engineering department at BMW, and his successor, Karlheinz Radermacher, shows no indication of making sweeping changes.

Quite independently of BMW's example, Harry Webster of Triumph developed a very similar system for the 2000 of 1963, in a step-by-step process of improving on the swing-axle design of the Herald. Webster chose a pivot angle of 28 degrees for the semi-trailing arms, added splines to the double-jointed shafts, and placed coil springs on top of the control arms.

It did not take long for the industry to catch on. Marcel Dangauthier of Peugeot chose semi-trailing arms for the 504, introduced in 1966, and Fred Hart designed a semi-trailing-arm rear end for the 1966 Ford Zephyr and Zodiac, with a 33-degree pivot angle. The following year, Joachim Sorsche designed a new rear suspension for the W-114 series Mercedes-Benz, using semi-trailing arms and coil springs in a layout that has since been adopted across the board for all Mercedes-Benz production cars.

Nissan/Datsun began using independent rear suspension with semi-trailing arms on the 1966 Bluebird/510, but reverted to live axles for the successors to these models about 1975. Volkswagen went to semi-trailing arms for the Beetle in 1967 and the 411 in 1969. Porsche did the same, starting with the 911.

In 1972 Ford of Europe went to semi-trailing arms and coil springs for its Consultaunus and Granada and built both in Britain and Germany. In 1978, Opel adopted semi-trailing arms and convex-profile coil springs for the Senator and Monza. There was no thought of getting variable rates, for the coils had uniform metal thickness. The reason was that the styling did not allow sufficient room for adequate spring length, and to avoid bottoming-contact between the coils with shorter spring height, chief engineer George Roberts chose to wind the coils loosely in the middle and tightly at the end, so that full compression would allow inter-coil overlap.

It is to be expected that Ford and Opel will extend the use of independent rear suspension to its lower-priced models and drive more of their competitors to do the same. Continued growth of front wheel drive on both sides of the Atlantic can only hasten the process.

Chapter 18
Evolution of Steering Gear

Correct steering geometry was a major problem for the pioneer car builders, for they had nowhere to look for answers except in their own inventive talent. They could not copy the front axle of the horse-drawn vehicle without accepting the inconveniences of fore-carriage steering (Fig. 18-1). A look at bicycles could show them one way of turning the wheels, but could not tell them anything about the proper relationship between the front wheel steering angles.

The principles attributed to Ackermann were unknown to most auto pioneers, even the most learned. And even the best technical universities had only four chairs in engineering: civil engineering, chemical engineering, electrical engineering and mechanical engineering. The task of the inventors of the automobile was made worse by the fact that there was no single or easily accessible source of information on the prior art in their field.

Ackermann's Principle Really Langensperger's

The Ackermann patent dates back to 1818, for instance. And when Amedee Bollee designed a steering linkage with correct Ackermann geometry for the 1873 L'Obeissante, he had no doubt arrived at it by himself rather than through instruction. Actually, Ackermann was not the inventor, but a vital middleman. The origins of Ackermann steering have been traced back to a carriage builder in Munich, Georg Langensperger, who obtain a Bavarian privilege for his invention on May 25, 1816.

Drawings from Langensperger's documents show an interesting comparison between the conventional fore-carriage type of vehicle and his new design with the split axle. With the old one, the whole front axle

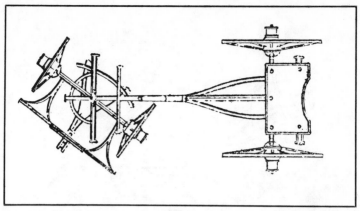

Fig. 18-1. The complete fore-carriage turned in order to steer horse-drawn vehicles, which placed both front wheels astride the same turn radius at right angles.

turned as a unit. Langensperger's wagon had a front axle with its center section fixed perpendicularly to the backbone frame. The wheels were mounted on stub axles (Fig. 18-2) linked to the main axle via elbow-shaped steering pivots, and steering arms extending diagonally backwards from the stub axles were connected to the main shackle bar via a track rod. This layout contains all the essentials of the Ackerman principle.

Rudolf Ackermann got into the picture because he met Langensperger in 1816 on a visit to Munich and was impressed with his front axle. Ackermann must be given great credit for realizing the value of what he saw. He was the son of a carriage builder, grew up in Saxony and apprenticed in his father's trade. He went to London before he was 25 and, after working for several coachbuilders, he started his own Printing shop. He made a fortune and donated large sums to aid the peoples of the German states after the Napoleonic wars. He agreed to act as Langensperger's patent agent in London, and after securing a British patent, he published a 60-page pamphlet about Improvements to Axle-Trees Applicable to Four-Wheeled Carriages.

Amedee Bollee (Fig. 18-3) used the same steering geometry for La Mancelle in 1878 (Fig. 18-4). In 1881, Charles Jeantaud used it for an electric car he was building. Jeantaud began digging deeper and more scientifically into the geometrical questions and formulating theories about it. As a result of his work, what we now call *Ackermann steering* became known in France as the *Jeantaud diagram*.

If Maybach made no use of the design when he built the Stahlradwagen in 1889, it is probably because he ignored its existence, and not because he disagreed with it. On this vehicle, the wheels made parallel turns. When Panhard-Levassor and Peugeot began to build cars, they used the Jeantaud diagram, and Benz patented a steering system of his own in 1893 that gave a similar effect (Fig. 18-5).

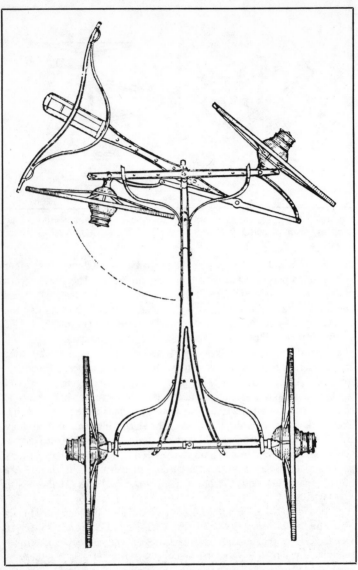

Fig. 18-2. Ackerman patented Langensperger's stub-axle steering, which also introduced new geometry. Wheels run on separate radii and must be turned at different steering angles.

Langensperger for Alignment, too

Aspects of wheel alignment such as camber and toe-in have been traced back to Langensperger. About 1800, the artillery wheel with wooden spokes came into use. The spokes were arranged in a dished

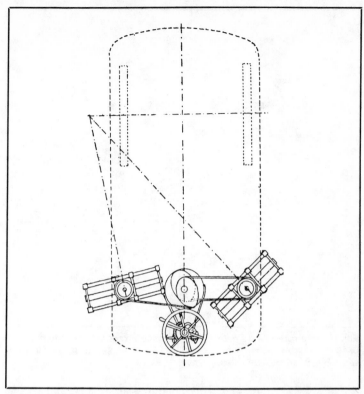

Fig. 18-3. L'Obeissante, built by Amedee Bollee in 1873, had Ackerman steering by a chain linkage to vertical steering swivels with fork-guidance for the front wheels.

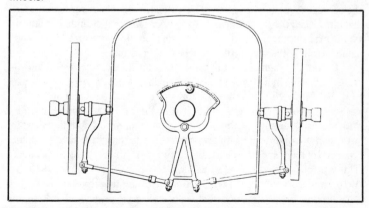

Fig. 18-4. La Mancelle, built in 1878, had steering arms parallel with the wheels, but a center link with diagonal intermediate arms that gave Ackermann steering geometry.

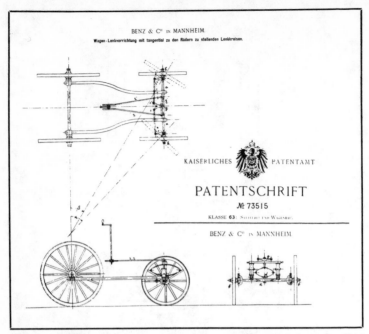

Fig. 18-5. Benz received a patent for his steering linkage with Ackermann effect used on Benz production cars since 1893. Tiller steering was used up to 1897, when Benz adopted the steering wheel.

pattern which greatly increased the lateral strength of the wheel. But their use necessitated pointing the tip of the stub axle down a few degrees so that the load on the wheel would always be borne by the spoke that was in the nearest-to-vertical position. This tilted the wheel outwards at the top and introduced the need for toe-in to compensate for the camber. Naturally, the first automobiles, which were little more than motorized carriages, came to follow the same rules of wheel alignmemt.

Carter also came in at the same time, for its effect was well known from the bicycle industry. Amedee Bollee does not seen to have made use of it, but Maybach tilted the twin forks of the Stahlradwagen at a rakish caster angle for directional stability. Later, Benz invented a purer system, with vertical king-pins mounted ahead of the front-wheel spindles so as to provide a positive trail, with the same benefits, and without raising the front end of the vehicle by means of the king-pins when turning the wheels. This type of caster was used by Vauxhall and Bugatti up to about 1920. By 1910, most cars obtained caster action by tilting the king-pin back.

King-pin inclination and center-point steering originated in Belgium shortly after the turn of the century. The 1902 F. N. Built to J. de Cosmo designs in the arsenal at Liege introduced this feature which was quickly understood and soon adopted by other car makers.

Eliminating kickback from ruts and bumps was a major difficulty, especially with tiller steering. The curved-dash Oldsmobile had a pin-joint in the middle of the cross-tie rod. This pin-joint had a short arm whose opposite end was splined to the bottom of the vertical tiller shaft. This gave an approximation of Ackervann steering, with a linkage that pre-

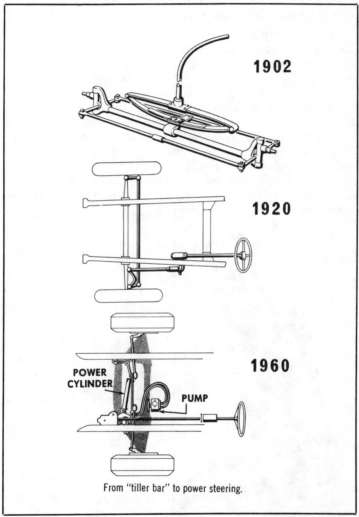

1902

1920

1960

POWER CYLINDER

PUMP

From "tiller bar" to power steering.

Fig. 18-6. The curved-dash Oldsmobile had a steering linkage with a single-piece drag link and adjustable tie rod in front of the axle. GM cars of the 1920's used worm-and-wheel steering and a simple fore-and-aft and cross-steer linkage. The parallelogram steering came in with independent front suspension, and power steering in the period from 1953 to 56.

vented road shock from knocking the tiller out of the driver's hands (Fig. 18-6).

Gradually the steering wheel replaced the tiller. Wheels for turning shafts were well known from the machine tool industry, and the Benz Velo of 1893 had a small wheel with a handle. By 1897, Benz had tilted the steering column back, enlarged the wheel and eliminated the handle. By 1900, most European cars had gone to wheel steering.

Rack-and-Pinion: Oldest Steering Gear

As for steering gear, it is rack-and-pinion that is probably the oldest. Leon Bollee used rack-and-pinion steering on his 1896 tricycle (with two steered wheels in front). Benz had dual steering racks on the 1893 Velo. And incredibly, Onesiphore Pecqueur's steam wagon of 1828 (which introduced the differential) also had rack-and-pinion steering.

Worm and sector, worm and nut, and rack and pinion steering gear systems were common by 1902. Worm and wheel steering originated at the Electric Motive Power company in 1902, an invention by P. W. Northey, and Henry Marles refined the worm-and-peg principle into the modern cam and roller steering gear in 1913.

The Marles gear consisted of two spiral cams fixed to the lower end of the steering shaft, bearing against a pair of rollers. These rollers were

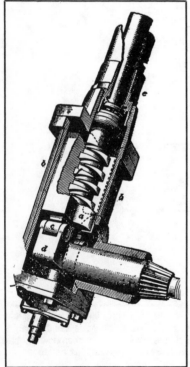

Fig. 18-7. Ross cam-and-lever steering gear was widely used on cars, as well as for trucks and buses, during the 1920's and 1930's.

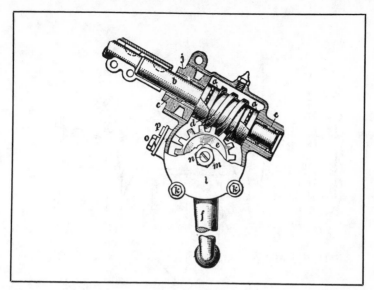

Fig. 18-8. Worm-and-wheel steering was the most popular type of steering gear on American cars from 1915 to 1925. This is a Dodge unit from about 1919.

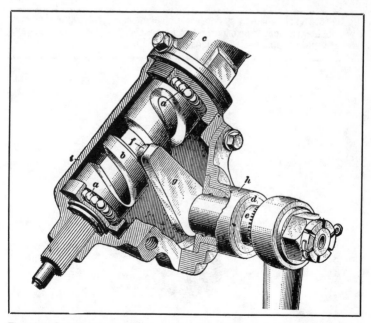

Fig. 18-9. Screw-and-nut steering was preferred by some car builders because of its low manual effort and absence of kickback. But if it was highly accurate, it also worked with high friction.

259

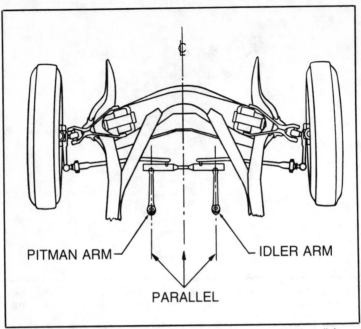

PITMAN ARM — IDLER ARM

PARALLEL

Fig. 18-10. The central relay system was one of the first parallelogram linkages to win popular application on either side of the Atlantic. It originated with the early independent-front-suspension designs.

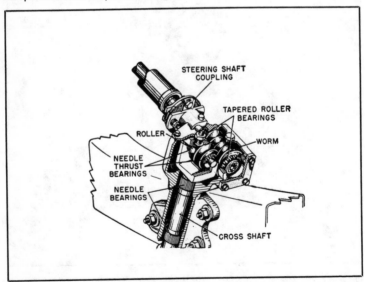

STEERING SHAFT COUPLING

TAPERED ROLLER BEARINGS

ROLLER

WORM

NEEDLE THRUST BEARINGS

NEEDLE BEARINGS

CROSS SHAFT

Fig. 18-11. Worm-and-roller steering works with low friction and can be made virtually free of reversibility. It is still in use for both cars and trucks.

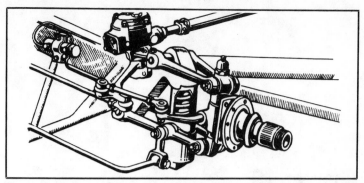

Fig. 18-12. Ferrari and Fiat were staunch adherents of worm-and-roller steering for their sports cars until quite recently. This is the steering gear on the 1953 Ferrari 250 GT.

mounted on a crank connected to the drop arm. Marles steering worked with purely rolling motion, which meant low friction, low wear, and no backlash.

R. Bishop got a patent for an improved type of cam and roller steering gear in 1923, with a single cam on the steering shaft engaging with a conical roller on a crank (Fig. 18-7). The cam groove was formed to fit the roller profile in all positions along its arc.

Recirculating-ball steering gear was developed by GM's Saginaw division between 1930 and 1940. What were they using before Buick had gone to worm-and-sector steering in 1909 and kept it till the 1930's (Fig. 18-8). Saginaw expeximented with hourglass-shape worms and roller-tooth gears as early as 1921 and Cadillac adopted this worv and wheel type steering gear in 1926.

Next, Saginaw brought out an hourglass-type worm and sector steeing gear for a 5-ton truck in 1929, with considerably greater helix angles than on earlier designs and a consequent increase in efficiency. By 1930, this was adopted by Oldsmobile, Cadillac and Pontiac. Buick had a new roller-tooth type of worm and sector gear, and Chevrolet switched in 1931 from its own type of worm and wheel gear to the new Saginaw worm and secator gear (Figs. 18-9 and 18-10).

Experimental work with ball-bearing roller-tooth gears began at Saginaw in 1933. The recirculating-ball-type of steering gear was first used on a few GMC trucks in 1939 and on the 1940 Cadillac V-12. All 1941-model Buicks and Cadillacs had it. Chevrolet and Pontiac continued using ball-bearing roller-tooth gears until 1955 and then adopted recirculating-ball steering gears (Figs. 18-11 and 18-12).

Starting in the early Fifties, Mercedes-Benz standardized recirculating-ball steering in all its cars. That seems to be accepted as the most efficient system for heavy cars (with and without power steering), while lighter cars-and some not so light, such as the Volvo 264 and Peugeot 604-use rack-and-pinion. Other types of steering gear may be about to disappear.

Chapter 19
Development of
Interconnected Springs

The idea of interconnected springs is to obtain self-equalizing suspension systems. That is a different concept from self-leveling suspension, though the two can be combined. Equalizing types of suspension work by means of a flexible link between the front and rear wheel assemblied, and their main task is to reduce pitching. They accomplish this by minimizing the difference between the vertical loads carried by the front and rear wheels.

Only one family of cars in production today has a form of equalizing suspension. It is the Citroen 2 CV/Dyane series of lightweight economy cars built on a short wheelbase of 94.5 inches. In this design, front and rear wheels are carried by bell-crank levers, leading at the front and trailing at the rear, pivoted in such a fashion that they move in vertical planes. A single horizontal coil spring assembly on each side of the car is connected to both front and rear suspension levers. The spring unit consists of two-coil springs enclosed in a cylinder, one facing forward, the other backward. In the initial production run, the springs were loaded in compression; but in 1955, the layout was modified so that the main springs could work in tension (the push rods from the suspension arms were simply converted to tie rods). Volute springs inserted between the cylinder and the tie-rod guides assured the requisite stability. The suspension linkages provided a roll axis at ground level, giving the car very low roll stiffness (Fig. 19-1).

That could easily be countered by fitting stabilizer bars; but in order to preserve the full wheel travel of these cars, which is exceptionally generous, and the same low spring rates, the Citroen engineers rejected the use of transverse interconnection (which is what the use of stabilizer bars would have amounted to). They are allowed to roll, resulting in

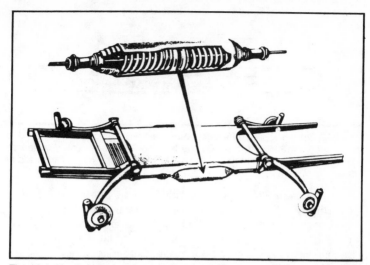

Fig. 19-1. Harris-Leon Laisne chassis from 1929 had all-independent suspension with interconnected front and rear coil springs placed horizontally within the tubular-frame side members.

camber angles that would be alarming on cars of higher performance levels.

Laisne Credited for 2 CV

The principles adopted for the 2 CV Citroen suspension were first drawn up and patented by J. J. Charley in 1910 and reinvented in 1927 by Leon Laisne (Fig. 19-2). They were used on the Harris-Leon Laisne cars, built in small numbers in Nantes up to 1931. These cars had an unusually low center of gravity, so that roll couples were kept within relatively minor values. But no real production enused.

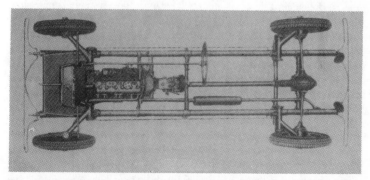

Fig. 19-2. Front wheels in the Harris-Leon Laisne were carried on leading arms with cranks to engage coil springs working some in tension and others in compression. Rear end had the same setup, but reversed to give trailing arms.

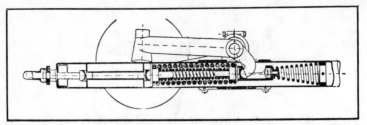

Fig. 19-3. Citroen applied for a patent on its centrally-supported interconnected suspension in 1934. On the 2 CV the front and rear wheels are carried on interchangeable arms, leading at front and trailing in the rear. Roll axis is at ground level.

On the basis of his experience with these cars , Leon Laisne formulated a theory on how and why interconnection of the front and rear springs on the same side of the car could substantially alter its response to road irregularities (Fig. 19-3). Later work in this field by other experts showed that fore-and-aft interconnection of road wheel springs could indeed reduce pitch frequency without lowering roll stiffness, consequently enabling engineers to select soft springs for good ride comfort without adversely affecting the handling characteristics of the car. Laisne protected his invention by a patent taken out in 1934, and two years later Dr. Frederick W. Lanchester presented a learned paper laying down the scientific rules of spring interconnection.

An experimental car with interconnected springs was built at Morris Motors in 1935, designed by Alec Issigonis and J. N. Morris. The car had all-independent suspension with lateral A-arms, front and rear, and longitudinal torsion bars. The torsion bars on each side were connected front-to-rear via a differential gear mounted on the frame midway along the wheelbase. But Morris never developed this ingenious design.

Citroen, on the other hand, adopted the Leon Laisne design for the small-car project that went on the drawing boards in 1935. Citroen boss Pierre Boulanger told his engineers he wanted an extremely simple, low cost machine, and to make his point, added, "Four wheels under an umbrella." He also insisted that the car should be so well sprung that it must be able to run, loaded with eggs, at speed across a plowed field and get to the other side with not one egg broken.

A prototype ran in 1936. Test and development work went on uninterrupted by by the war, for Boulanger apparently convinced the German occupation forces that they were working on a revolutionary type of military vehicle-an interesting project from Wehrmacht's point of view.

What became the 2 CV was ready in 1946 (Fig. 19-4). All things considered, tooling-up for production was remarkably quick, and the car was presented to the public at the Paris auto show in October, 1948. When the show closed, Citroen had orders for three years' production! And the same basic car is still being built, 31 years later, at rates exceeding 200,000 a year.

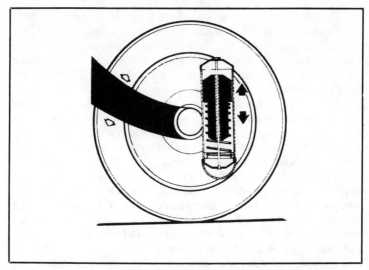

Fig. 19-4. Instead of hydraulic- or friction-type shock absorbers, the 2 CV Citroen wheels carry harmonic dampers which tend to check a resonant frequency of oscillation without interfering with initial wheel travel.

Packard's Torsion Level Ride

Packard, unlike Citroen, did not design a completely new car for their first model with interconnected springs. Packard used an existing chassis and merely replaced the standard coil springs with a new torsion bar system invented by William Allison. The wheel locating members needed only minor changes to accomplish this, and the Torsion Level Ride was introduced in the 1956 Packards (Fig. 19-5).

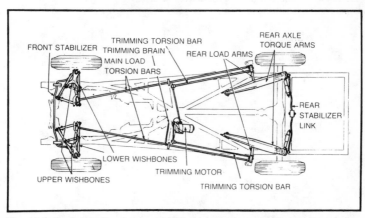

Fig. 19-5. Packard's Torsion Level Ride had front/rear interconnection via torsion bars. Additional trimming torsion bars for the rear suspension provided automatic level control.

The front suspension system used short and long transverse triangular arms, and the rear axle was located by radius arms. Torsion bars carrying the main load of both front and rear masses ran alongside the frame side members, and a shorter set of parallel torsion bars worked as a leveling device for the rear, assuring freedom from pitch. The ends of the short torsion bars were connected to an electric motor by levers. The motor was activated by load variations, and the torsion bars maintained the body in a permanently horizontal position regardless of load and weight distribution. The control system worked through a relay giving a six-second time lag between input and activation so as to avoid reaction on transient deflections. Six seconds after receiving a heavy suitcase in the trunk, the torsion bars would re-elevate the tail of the car to its normal height. The car was sensitive to weight transfer on braking and acceleration, however, and it needed stabilizer bars, both front and rear, to limit lateral weight transfer. From the passengers' viewpoint, the car had improved ride qualities because the natural pitch frequency was lowered. The Torsion Level Ride was discontinued, however, when Studebaker brought Packard production to South Bend, and all experimentation in this area ceased.

About 1964, Ford of England patented a system of mechanical spring interconnection which is reminiscent of the Laisne and Citroen 2 CV principles. But Ford introduced a new element which may well be essential to the future success of mechanical spring interconnection-a front ear attitude spring which remains almost inactive on single-wheel bumps, but comes into effect when both front or both rear wheels are deflected, thereby causing an increase in the overall spring rate. It is connected to the front and rear suspension springs and works both ways, i.e. it can both add to and detract from the force of spring actions. However, Ford has made no use of this invention in its production cars.

Chapter 20
From Snubbers to Gas Dampers

Prior to 1900, automobiles had little need for shock absorbers. Springs were stiff, and the vehicles light. Drivers and passengers accepted the jolts and the harshness in return for greater speed, for horse-drawn carriages usually had a pleasant ride, but were restricted to equestrian velocity. As cars were built with softer springs, the need for a spring damper became more obvious. The friction-type shock absorber was already in existence, but its inventor had little idea of the potential demand for it.

The inventor was Jules Truffault, an absent-minded bicycle builder in Paris, whose hollow front fork and other inventions had contributed greatly to the progress of the racing bicycle. In 1899, he adapted this fork to motorcycles and motor tricycles, and it was noticed by an American business man, Edward V. Hartford, who had earlier tried (and failed) to set up a U.S. factory for making de Dion-Bouton motor tricycles. He met Truffault and bought bicycle forks and other parts from him.

Franco-American Development

One day, Truffault showed him the shock absorber. It consisted fo two friction discs attached to separate arms pivoted on the same axis. The arms described a scissors-like action on wheel deflection, one arm being attached to the axle and the other to the chassis frame. One disc was made of oil-soaked rawhide, anf the other of bronze. The pressure on them was provided by fastening nuts with large spring washers. Thus, they were adjustable and, in fact, needed periodic adjustment (Fig. 20-1).

Hartford went back to the U.S. in the autumn of 1900 and sent an Oldsmobile to Truffault for him to develop a shock absorber for au-

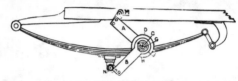
Fig. 20-1. A 1904 advertisement for the Truffault-Hartford shock absorber clearly showed its mode of action, and quoted races won with Fiat, Mercedes, Gobron-Brilie and Richard-Braiser cars equipped with these friction dampers.

tomobiles. Hartford acquired the American rights to it and started production in 1903. Truffault made license agreements with French car makers to build their own, and by 1904, his shock absorbers were used by Richard-Braiser, Mors and Peugeot.

The problem with Truffault's basic principle was that the mechanism could not distinguish between the initial bump and the subsequent spring flexing. It offered full resistance to all wheel travel. Its action also had an element of unpredictability. It would stick more on a cold morning than on a hot afternoon, for instance. De Ram, Viet and other tried to overcome this drawback by using multi-plate assemblies and improving the control of the spring-load on the friction plates. Others turned to different basic principles.

Hydraulic Concept Came Early

The notion of obtaining the desired resistance by passing a hydraulic fluid through a restriction occurred to several inventors at a remarkably early date. Mors in Paris built a car equipped with single-acting pot-type hydraulic dampers in 1902.

The telescopic damper was patented as early as 1901 by C. L. Horock, using hydraulic fluid, a cylinder and a piston with a non-return valve-anticipating in almost exact detail the modern shock absorber. In 1905, Louis Renault invented two different types of hydraulic shock absorbers, one with a single lever actuating two pistons placed at opposite ends of an oil-filled cylinder, and the other a telescopic-type which was a direct forerunner of the type now fitted on the majority of cars. Renault used the piston-type hydraulic shock absorbers on the Grand Prix car of 1906, but, he did not put it on his mass-produced cars.

The shock absorber, whether friction-type or hydraulic, was destined to remain an aftermarket item for many years. One type reached such popularity in America and abroad that its trade name became almost synonymous with the product: The "Gabriel snubber." (See Fig. 20-2). It was invented in 1915 by Claud Foster, an auto dealer in Cleveland who also

Fig. 20-2. The Gabriel snubber was a rebound-control device that relied for its action on a spring-loaded strap wound around a block.

Fig. 20-3. The Houdaille damper worked with hydraulic vanes pumping through restrictions between the internal chambers. It was double-acting and air-cooled.

ran a manufacturing company, Gabriel. It was called "Gabriel" and not "Foster" because its first product was Gabriel's Horn, an audible warning device using exhaust gas. Foster was also a trombone player, and he got the idea from his musical instrument.

The Gabriel snubber consisted of a flexible strap wound around two movable blocks separated by a compression spring. It offered no jounce resistance, but it slowed down the rebound as the strap tightened around its blocks. For some years, the Gabriel snubber accounted for 75 percent of the shock-absorber business in America.

Houdaille Vane-Type Displaces Snubber

Hydraulic-types were soon to make a challenge that would completely kill the snubber market, however, this challenge came from the vane-type hydraulic damper invented by Maurice Houdaille, who began to develop the hydraulic damper about 1906 and became the world's leading manufacturer of shock absorbers by 1910 (Fig. 20-3).

Houdaille's type worked with two arms, one attached to the axle and the other to the chassis frame, mounted on the same pivot shaft. This shaft was located inside a cylindrical container where each arm carried vanes working against a volume of hydraulic fluid. First to use the Houdaille damper as original equipment was Rochet-Schneider of Lyon in 1909. Shortly afterwards, Houdaille dampers became standard equipment on certain models of Benz, Vauxhall, Sunbeam and Panhard-Levassor.

The first Houdaille dampers were single-acting, being effective only on rebound. But about 1921 a new valve system was adopted to give damping also in jounce, making it truly double-acting. Chenard-Walcker, Delage, Farman and Unic began fitting the Houdaille dampers.

Piston-Types Led to Telescopics

About 1930 an American branch was opened in Detroit to supply both the industry and the aftermarket with shock absorbers. In the meantime, General Motors had gone into shock-absorber manufacturing in a big way. The Delco design was based on a patent from 1907 for a piston-type

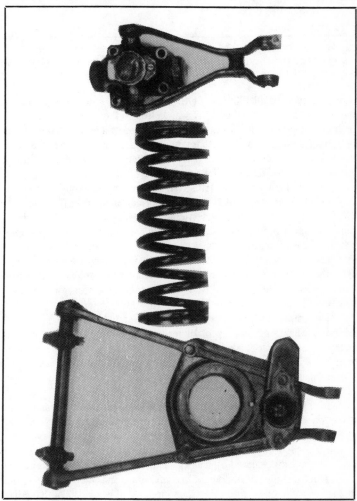

Fig. 20-4. Hydraulic piston-type shock absorber (top) made by Delco for the 1936 Oldsmobile. Its levers are formed by the upper control arm. For size comparison, the lower arm and coil spring are also shown.

271

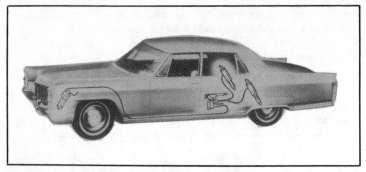

Fig. 20-5. 1965 Cadillac with Delco Superlift shock absorbers on the rear axle, compressed air reservoir provided boost to gas-filled cells on top of the hydraulic section in order to provide some measure of level control.

designed by C. Caille. In this unit, a single lever followed suspension movement, and the damper body was bolted to the chassis frame. The body contained two parallel cylinders connected by a valve system at the bottom. The top end of the pistons contacted cams moved by a rocker arm carried on the lever pivot shaft. It remained in use at GM until the Fifties but was finally replaced by the telescopic-type (Fig. 20-4).

T. B. Andre and Gordon Armstrong built piston-type shock absorbers in England for many years and became important suppliers to the industry. Gordon Armstrong started production of shock absorbers in 1921 and got his first factory orders in 1928. In 1930, he patented a telescopic damper which won immediate acceptance.

The early development of telescopic shock absorbers stems mainly from the work of Vincenzo Lancia, who incorporated double-acting hydraulic units in the sliding-pillar front suspension system of the 1923 Lambde.

The same idea came to the engineers of the Brisk Blast Co. of Monroe, Michigan, by way of their experience with pneumatic hand pumps. But its first production dampers were piston-types, resembling the Delco. In 1933, the business had become the Monroe Auto Equipment Co. and introduced the telescopic shock absorber in America. Hudson was the first original-equipment customer.

During the late Forties, Adrien G. de Koning of Oud-Beijerland in the Netherlands perfected a new type of telescopic shock absorber-one that was adjustable for hardness and could be rebuilt (the conventional type was discarded when worn-out). That became the Koni shock absorber which was very quickly adopted by leading manufacturers of high-performance cars, found a ready aftermarket with professional and amateur racers, and gradually spread into auto-enthusiast circles and beyond. The Koni has been widely copied by other companies.

At the same time, Christian Bourcier de Carbon, a French mining engineer and specialist on studies of tire adhesion to the roadway and the influence

272

of wheel travel on adhesion, developed the theory of the gas-filled shock absorber. He founded the Societe des Amortisseurs de Carbon about 1950 and became a manufacturer. Hans Bilstein, an auto accessory manufacturer in Germany, bought the rights to the de Carbon patents and began building gas-filled shock absorbers in 1958. Other major companies followed within the next 20 years (Fig. 20-5).

This sums up the progress up to the present state-of-the-art as far as shock absorbers for suspensions systems with metallic springs are concerned. Beyond that, one must look to self-leveling suspension units.

Chapter 21
Air/Gas/Liquid
Springing Development

Obtaining elasticity in vehicle suspension systems by pneumatic means is an idea that antedates the gasoline automobile by a considerable margin of time. A British patent for air springs intended for horse-drawn vehicles was taken out by Thomas Walker and George Mills in 1845, and American wagon builders tried air springs on experimental carriages during the Civil War.

One of the first automobile applications of an auxiliary air spring was the French-made Aeros—a ballon installed between the leaf spring and frame-sold as an aftermarket accessory in 1905. The American Twombly car of 1906 had a similar device. The Aeros and Towmbly soon disappeared, but the idea was revived and refined by another French inventor, R. Bernat, in 1920. It used a low-pressure bellows and was primarily intended for trucks. A. Sharp patented a piston-and-cylinder type of air spring in England in 1906, but there is no record of its use.

The Kilgore automobile air cushion, made in Boston, came on the market in 1905. It consisted of two bronze cylinders telescoping into each other. Both were sealed on top so that the smaller-diameter one worked as a piston inside the larger cylinder. The working air was trapped between them. In 1909, Mellen-Edwards of Beloit, Wisconsin, offered a development of the same basic idea, with two air cushions per wheel attached to the axle at the bottom and splayed to mounting points on the frame at their upper ends.

First Real Systems in 1909

All of these systems were purely auxiliary, as the standard leaf springs were retained. The first true air-spring systems were the Cowey (from England) and Delpeuch (from France) in 1909. Both were tested by a number of auto manufacturers at the time, but neither made it into production.

The Cowey system, patented by L. E. Cowey in 1908, comprised an engine-driven compressor and a spring unit at each wheel. The spring units were vertical cylinders with pistons carried on axle-mounted rods. On top of the cylinders were spherical chambers functioning as reservoirs, separated from the cylinders by valves. The Delpeuch system was similar in principle, but differed in its execution in that the spring units lacked the individual reservoirs on top. A patent from 1909 by F. de Lostalot and M. Guerpillion described a system of air springs with height-adjusting valves connected to the steering column.

First Major Auxiliary Air Spring

The first large-scale commercial application of auxiliary air springs was the Westinghouse unit, an invention of George Westinghouse, made in Pittsburgh, Pa., by the Westinghouse Air Spring Company for 1912 until about 1924. This air spring was a vertical cylinder used in conjunction with leaf-spring chassis. The cylinder was interposed between one end of the leaf spring and the frame (as a forward shackle on the front springs).

The cylinder contained a piston with an air chamber on top and was filled with oil to the piston crown. The piston was a complex assembly with plungers and valves to control air and oil flow. The Westinghouse air spring was never adopted as original equipment, but it enjoyed a certain popularity in the aftermarket. It disappeared when front wheel brakes came in (probably due to its inability to function as intended under conditions of high thrust loads in the front suspension system).

The success of the Westinghouse device inspired many imitators, such as the Air-O-Flex of 1917, Gruss of 1922 and Thompson of 1924. The Pneumobile Six was announced in 1915 as a product of the Cowles-McDowell Pneumobile Company of Chicago, Illinois, but never progressed beyond the prototype stage. The Pneumobile system relied on vertical cylinders with pistons compressing the air in the upper end by rods attached to axle. In principle, it was similar to the Cowey and Delpeuch devices. Pneumatic cylinders were installed in forks formed in the chassis frame.

Messier: Significant Progress and Production

The next important step of progress came in 1922 with the Messier car, designed by Georges Messier. This young French engineer had begun to develop his basic idea in 1920 and installed experimental units on a number of test cars. The Messier system was no more advanced in its basic concept than its predecessors, but it was thoroughly engineered and enjoyed full reliability. Messier used an engine-driven compressor feeding into a central reservoir and mixed a certain amount of oil with the compressed air to assure better sealing. The spring units were equipped with

Fig. 21-1. Messier prototype from 1927 used air springs as the sole weight-supporting and elastic element. Front axle was located by radius rods and a track bar.

non-return valves that sensed vertical loads and admitted more air into the chambers when load was added, so that the chassis was always level and maintained constant ground clearance. (Figs. 21-1 and 21-2).

The Messier car was in production from 1922 to 1930. His patents form the basis for most subsequent developments, not only for automobile suspension systems but also for aircraft landing gear. It was due to the success of his olco-pneumatic shock absorber and his hydraulically retracting landing gear that he got out of the auto business. Georges Messier died in 1933, but his chief engineer, Ing. Breguet, carried on the research and

Fig. 21-2. Messier rear axle had air-spring cylinders placed outboard of the frame, with trailing arms and a short (but hefty) track bar.

276

development work. The Messier company played a part in the early development of the Citroes suspension (as recounted later) and is very much alive today, though its main customers are not automotive, but aerospacial.

No doubt inspired by the early Messier experiments, engineers of the Beardmore works in Scotland went to work on a different system, which was first shown in 1923. The principles were the same as in Messier's case. However, Beardmore never developed the system; several experimental cars were built, but it was never placed in production.

At Beardmore and elsewhere, engineers soon saw that air, which is a compressible gas, was an excellent medium for supporting the load. But air springs needed damping even more than metallic springs. With pneumatic springs, it seemed incongrous to use the crude friction-type shock absorbers, especially since hydraulic ones existed.

Hydro-Pneumatic and Other Attempts

Liquids are imcompressible and are therefore, first-class damping media. This line of thought led to the air-and-oil, or hydro-pneumatic, combination.

The first attempt along these lines was made by Louis Renault, as shown in a 1908 patent. Both front and rear axles were located by trailing arms,and each wheel hub was connected via a vertical pushrod and a bell-crank to the piston shaft in the air-filled suspension cylinders. Each piston shaft was connected to two pistons, one large and one small, working against liquid-filled chambers in opposite ends of the cylinder. The large pistons supported the static weight of the car, and the small pistons dampened the wheel movements. Both liquid chambers of each cylinder were interconnected, and tubes connected all cylinders to a central reservoir. The reservoir was pressurized by an air pump operated manually.

Presumably Renault built prototypes with this suspension system, but they never reached the production stage, and the idea lay dormant until machine tool engineering had developed sufficiently to ensure the precision needed for reliable operation over long periods of time without attention.

An Italian system, resembling Beardmore's, was built and tested by Carlo Rigotti, who installed it on a Fiat 509. Rigotti used rubber air-bellows placed inside coil springs. The system is said to have failed because of poor quality in the fabric and rubber used.

Another French pioneer of air suspension elements was Jean Mercier, who installed air springs on a car in 1928 (Fig. 21-3). His system was unique in that the air was compressed to 735 psi. This was needlessly high for a passenger-car installation and may have been the reason for its downfall. Later, Mercier succeeded in developing an air-cushion shock absorber for aircraft, and the AER shock absorber went into production in May, 1935.

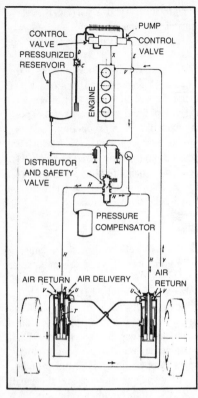

Fig. 21-3. Mercier air-suspension system used an engine-driven compressor, a reservoir and a compensating tank. Pressure in the spring units was controlled by a central air distributor and safety valve.

Firestone and the Air-Sprung Bus

About 1930, Firestone began an air-spring program. It was headed by its chief research engineer, Roy W. Brown. The result was a new flexible air bellows (looking like two giant doughnuts, one on top of the other) with individual air-pressure controls to each of the four units (Fig. 21-4). A pendulum valve directed the control units, so that the system automatically compensated not only for bump deflections, but also for body roll and nose dive.

The Firestone spring units were interchangeable with the coil spring in the new independent front suspension systems appearing on many GM and Chrysler 1934 models, so that installation called for no mechanical modifications. For rear installation, considerable reworking of the standard suspension was necessary. (Fig. 21-5).

Firestone converted two cars in 1935-a 1935 Buick and a 1934 Plymouth. Both were demonstrated in Akron, Detroit and elsewhere. Over the following years, Firestone made successful installations on Studebaker, Chrysler, Cord, LaSalle, Lincoln and Checker cars; and William B. Stout chose the Firestone air springs for his 1935 Scarab.

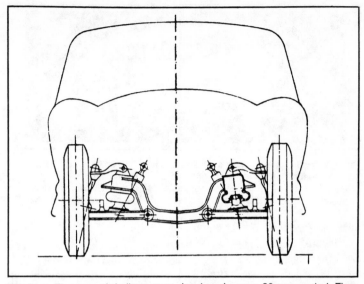

Fig. 21-4. Firestone air bellows were developed over a 20-year period. Firestone's proposal from 1957 indicates exactly what GM was to do in 1958-59.

Firestone was given new impetus in 1938 when General Motors became interested in using its system on buses. The first buses so equipped were road-tested in 1944, but reliability was a problem. It was not until 1953 that GMC Truck and Coach Division placed the air-spring bus in production, but it is important to note that when this step was taken, air springs completely replaced metal springs for GMC buses.

Fig. 21-5. Firestone design for air-spring installation on the live rear axle of a car with four-link rear suspension.

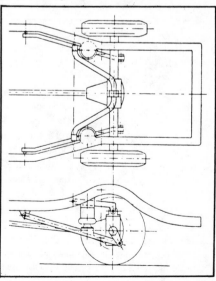

Fig. 21-6. Hydro-pneumatic suspension was standard on the Citroen DS-19 of 1955. Suspension units (1) are connected to a high-pressure cylinder (2) via a pressure control valve (3) and accumulator (4). The distributor (5) balances front and rear in respect of load and the assist of a brake-proportioning valve (7).

Spectacular Citroen, Failure at GM

It is also important to keep the bus experience in mind when we get to Cadillac's later use of air springs. One was not the result of the other, however, but prompted entirely by outside events. I am referring, of course, to the announcement of the Citroen DS-19 in October, 1955, with its standard oleo-pneumatic suspension system (Fig. 21-6 and 21-7). A rear-wheels-only version had been introduced on the six-cylinder 15 CV models in 1954, but it never had much impact.

Air-spring experience at Citroen began in 1939 when 40 s were equipped with a new system under development by Messier. Testing went on throughout the war period and into the postwar era. As development work progressed, the system became more and more Citroen and less and

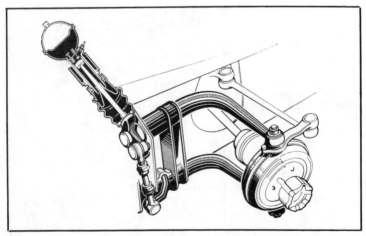

Fig. 21-7. DS-19 front wheels were located by lateral control arms of equal length. Spring steel elbows provided some horizontal compliance.

less Messier. If any one man is to be credited with the results, it is Paul Mages, who is still at Citroen, now as co-director of the aggregate division of the engineering department.

Technically, the DS-19 with its centrally-controlledcontrolled hydro-pneumatic suspension was a bombshell. Seismic shock was felt intensely in Detroit, and especially at General Motors, which was in the process of building its new technical center at Warren, Michigan, and promoting its research and engineering activity under the slogan, "The Spirit of the Inquiring Mind." Suddenly a car came from Paris which showed that French minds had not only inquired, but had found answers that were at once radical, logical and complete.

More or less as a direct consequence, air springs became a necessity for Cadillac, at the earliest possible date. A system was in the works at The GM Engineering Staff, where Von D. Polhemus and Lawrence J. Kehoe, Jr., had not only proved the unsuitability of a scaled-down version of the bus system for passenger-car installation, but had also proposed a basic air-spring system designed specifically for cars. Developmental work was accelerated to get it ready for production, and Cadillac engineers Fred H. Cowin and Lester Milliken were given the task of adapting the system for installation on the 1957 Eldorado Brougham.

The Cadillac spring units consisted of an air dome with a diaphragm. The piston acted on the diaphragm and was connected to the lower control arm. For 1958, the system was adapted to other Cadillacs and was made available to the other GM car divisions. Cadillac standarized it on the Seville and Biarritz models. Buick used it for rear wheels only, while Oldsmobile, Pontiac and Chevrolet adapted Cadillac's four-wheel system. By the end of the 1959-model year, the air-spring option was discontinued. Why? Reliability was unsatisfactory, to say the least. Price as another deterrent ($215 of the Cadillac) to popularity.

A superior suspension system was developed for the Firebird III experimental gas turbine prototype in 1959. Here, the chassis group led by Joseph Bidwell switched from pure air springs to an oleo-pneumatic system with interconnection, all much closer to the Citroen principles than the GM production system.

As in the DS-19 system, a rubber diaphragm separated the oil-based and gas-based portions of the spherical spring containers. A piston transmitting suspension movement led into each spherical container, and the accumulator body included damper valving to function in lieu of shock absorbers. Two such units were used at each rear wheel; the front wheels had one unit each.

The interconnection consisted of hydraulic transfer pipes between the front unit on each side and the forward unit in the rear suspension on the same side. This form of interconnection reduced the single-wheel spring rate while maintaining a reasonable roll rate. It also reduced the pitch stiffness of the springing to some extent.

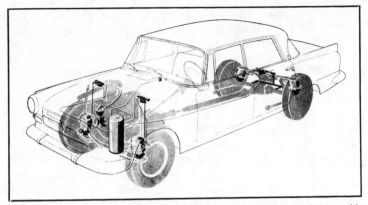

Fig. 21-8. Borgward and Continental Gummiwerke collaborated to produce this all-pneumatic suspension system for the Borgward 2300 in 1959. The car had all-independent suspension.

Many Other Attempts

In the meantime, Ford and Chrysler had also had their experience with air springs. Ford got into it gradually. The 1957 Mercury used an auxiliary air cushion as the forward shackle on the rear leaf springs. It was more a matter of filtering out vibration than of handling springing functions. But on the 1959 Mercury, a four-wheel system of true air springs was offered. Both designs came from the fertile drawing board of George Muller, a prolific inventor who is still at Ford.

Fig. 21-9. Air-spring unit of the Borgward 2300 was positioned on the lower control arm and abutted in a dome in the front crossmember. Hydraulic shock absorbers were mounted on the upper control arms.

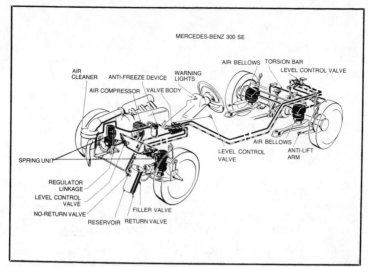

Fig. 21-10. A high-pressure-air-suspension system was used on the Mercedes-Benz 300 SE in 1961, providing automatic level control and constant spring rates regardless of load.

Chrysler did not develop a four-wheel system, but made rear-axle springs optional on the 1958 Imperial. As in the case of GM, it was all over by 1960. American Motors also offered air springs in 1958, but only for the rear wheels and only on the Rambler Ambassador.

All the favorable publicity that greeted the air-sprung system had its impact in Europe, too. Firestone offered a new passenger car system to

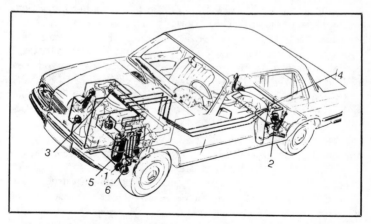

Fig. 21-11. Mercedes-Benz switched to hydro-pneumatic suspension for the 450 SEL 6.9 in 1975, in a general adoption of Citroen's principles. (1) and (3) are spring units, (6) and (7) the front and rear level regulators, (11) the main reservoir and (12) the central reservoir.

Fig. 21-12. Hydrolastic suspension introduced on the Morris 1100 in 1962 was a combination of rubber springs with hydraulic front/rear suspension.

Renault had a brief romance with auxiliary air springs for the Dauphine and Caravelle models of 1956 to 62 (also relying on Messier technology).

Lockheed had developed an oleo-pneumatic strut, primarily intended for aircraft applications, which was incorporated in the suspension system fo the fabulous V-16 BRM racing car of 1950. The strut weighed only four the British auto industry, only to find that both Dunlop and Armstrong had also developed similar systems. Firestone (who then had a controlling interest in Phoenix Gummiwerke of Hamburg-Harburg) next looked for customers in Germany. And miraculously, they found one. Borgward of Bremen worked closely with Firestone-Phoenix and Bosch to develop an air-spring system for its 1959 model 2400 and rushed it into production. Within 18 months of its debut, Borgward was bankrupt, so the car never got much chance to prove itself or the air springs. (Fig. 21-8 and 21-9). pounds and did not take more space that a normal shock absorber; but it did not lead to further use on racing cars, and no production cars followed BRM's lead.

Some years later, Toyota adopted pneumatic spring units for the front suspension on the Century V-8 series, which remains in limited production.

Mercedes-Benz engineers Nallinger, Scherenberg, Sorche, and Uhlenhaut began experimenting with air springs about 1955, and the resulting system was introduced late in 1961 for the new 300 SE sedan (Fig. 21-10). It was a pure air system, as opposed to Citroen's and depended on large-size, special De Carbon hydraulic shock absorbers for damping action. Its high cost was its bane. Mercedes maintained it in low-volume production until the models 300 SE, and 300 SEL 4.5 and 6.3 we re discontinued in 1972. When the 450 SEL 6.9 was introduced in 1975, it was equipped with a new hydro-pneumatic system (Fig. 21-11).

Citroen continued to refine its system and used it for the SM of 1970 and the GS of 1971 (which is still being made at the rate of 250,000 cars a year). The latest version, described earlier, is used on the CX which came out in 1974.

Hydrolastic Derivation

A sideline from pneumatic suspension was the Hydrolastic, a rubber-and-liquid system with front ear interconnection developed by Alexander Moulton and Alec Issigonis (Fig. 21-12). First presented at Earls Court in 1962 as a feature of the Austin and Morris 1100's, it was also used on the Mini from 1964 to 1969 (replacing the pure-rubber suspension system), the Austin 1800 introduced in 1964, and the Austin Maxi of 1969.

While Citroen used nitrogen gas and a liquid closely related to brake fluid, Moulton chose to substitute rubber for the gas and a combination of water and alcohol (to prevent freezing) for the oil-based liquid. Hydrolastic is self-damping, without separate shock absorbers.

One Hydrolastic unit was placed in connection with the locating arms of each wheel, the unit consisting of a cylindrical metal casing containing a conical rubber spring. The casing was fixed to the frame of the car, and its

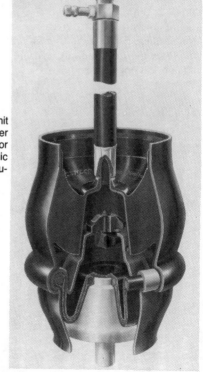

Fig. 21-13. Hydrolastic spring unit was delevoped by Moulton Rubber Co. and produced by Dunlop for British Motor Corporation. Hydraulic fluid gave anti-dive effect and automatic level control.

bottom wrapped around so as to form a hydraulic seal with a pressed steel spacer which had a bleed valve in its periphery and large ports in its top surface. These orifices were sealed by flexible-rubber flap valves on top and bottom, and worked as shock absorbers. Below this metal casing was a flexible rubber diaphragm, nylon-reinforced, which sealed the lower end of the chamber. The diaphragm was moved by a tapered piston connected to a suspension control arm.

The Hydrolastic chamber was filled with the water-and-alcohol mixture. When the piston rod was driven upward by jounce travel, the diaphragm flexed and displaced water through the damper restriction. As the volume of the cylinder was reduced by the motion of the piston, the rubber was correspondingly deformed by the pressure of the liquid.

The cylinder, diaphragm and rubber springs were so shaped that the volume of liquid displaced increased at a higher rate than the linear increase in piston stroke, with the result that the rubber spring offered a progressively higher resistance to load. The use of water instead of an oil-based liquid was dictated by the constant viscosity of water at all normal operating temperatures.

By connecting the front and rear Hydrolastic units on each side of the car, further advantages were gained: virtual elimination of pitch movements and balanced distribution of load on the outside wheels during cornering (Fig. 21-13).

Hydrolastic has since given way to Hydragas, introduced in 1973 and described earlier. It is the Citroen principles that have been best vindicated over the years and will, no doubt, find more widespread application in future cars of all types.

Appendix

The following three charts list design family groups for independent front suspension, independent rear suspension, and rear axle suspension. Use them for quick reference.

Table A-1. Design Families of Independent Front Suspension

SPRING TYPE	Sliding pillar	Twin I-beam	Trailing links	De Dion axle	Dubonnet	Equal-Length A-Arms	Short & long A-Arms	MacPherson
Semi-elliptic leaf springs		1931 Unic		1897 Graf				
Transverse leaf	1897 Decauville		1935 Alfa Romeo			1878 A. Bollee	1931 Peugeot	
Coil springs	1904 Christie	1965 Ford F-100	Aston-Martin Atom		1934 Chevrolet & Pontiac		1932 Mercedes-Benz 380	1950 Ford Consul & Zephyr
Torsion bars			1933 Mathis Emysix		1948 Vauxhall Velox & Wyvern	1935 MG R-Type	Citroen 7CV	1969 Fiat 130
Rubber							1936 Lightweight Special	
Air							1958 Cadillac	
Air & Oil			1950 BRM			1955 Citroen DS-19		1969 Porsche 911

Table A-2. Design Families of Independent Rear Suspension

SPRING TYPE	Swing axles	de Dion axle	Equal-length A-Arms	Short & long A-Arms	Trailing arms	Semi-trailing arms	MacPherson
Semi-elliptics leaf springs	1903 Adler	1896 de Dion-Bouton	1923 San Gius	1922 Sizaire Freres			
Transverse leaf	1920 Tatra	1937 Delahaye	1927 Cottin-Desgouttes	1922 Sixaire Freres	1937 Lancia Aprilia		1969 Fiat 128
Coil springs	1931 Mercedes-Benz 170	1939 Mercedes-Benz 770		1960 Lotus F-2	1939 Atalanta	1950 Lancia Aurelia	1959 Lotus Elite
Torsion bars	1933 Mathis Emysix	1937 Mercedes Benz W-125	1935 MG r-Type	1946 Lagonda	1934 Citroen 7 CV		
Rubber				1945 Alta	1959 Austin Mini		

Table A-3. Design Families of Rear Axle Suspension

SPRING TYPE	Chain drive and radius rods	Shaft drive and radius rods	Hotchkiss drive	Torque-tube drive	3-link and 4-link system
Full-elliptics	1885 Benz	1898 Renault		1910 Buick	
Semi-elliptics	1892 Panhard-Levassor		1905 Hotchkiss	1902 Excelsior	
Quarter elliptics				1922 Austin	1921 Leyland
¾ elliptics		1907 Cadillac		1913 Overland	
Platform springs	1907 Apperson	1905 Rolls-Royce			
Cantilever springs	1901 Olds	1911 Rolls-Royce		1916 Fergus	1903 Lanchester
Transverse leaf springs				1906 Ford	1954 Ferrari
Coil springs	1894 Daimler			1937 Buick	1950 Aston Martin
Torsion bars					1947 Frazer-Nash

Index

290